Membrane Protein Models

Membrane Protein Models

J.B.C. Findlay
Department of Biochemistry and Molecular Biology,
University of Leeds, Leeds, UK

First published 1996

A CIP catalogue record for this book is available from the British Library.

ISBN 1 85996 080 4

BIOS Scientific Publishers Ltd
9 Newtec Place, Magdalen Road, Oxford OX4 1RE, UK
Tel. +44 (0)1865 726286. Fax +44 (0)1865 246823

DISTRIBUTORS

Australia and New Zealand
DA Information Services
648 Whitehorse Road, Mitcham
Victoria 3132

India
Viva Books Private Limited
4325/3 Ansari Road, Daryaganj
New Delhi 110002

Singapore and South East Asia
Toppan Company (S) PTE Ltd
38 Liu Fang Road, Jurong
Singapore 2262

USA and Canada
Books International Inc.
PO Box 605, Herndon, V A 22070

The colour section in this book is kindly sponsored by Orion-Farmos

Typeset by Euroset, Alresford, Hampshire, UK.
Printed by Information Press Ltd, Oxford, UK.

Contents

Contributors

Argos, P. European Molecular Biology Laboratory, Postfach 102209, Meyerhofstrasse 1, D-69012 Heidelberg. Germany

Bhogal, N. Department of Biochemistry and Molecular Biology, University of Leeds, Leeds LS2 9JT, UK

Blaney, F.E. SmithKline Beecham Pharmaceuticals, Computational Chemistry Department, New Frontiers Science Park (North), Third Avenue, Harlow, Essex CM19 5AW, UK

Bradshaw, C. Glaxo Institute for Molecular Biology, Plan-les-Quates, CH-1228 Geneva, Switzerland

Chollet, A. Glaxo Institute for Molecular Biology, Plan-les-Quates, CH-1228 Geneva, Switzerland

Cocchi, M. Dipartimento di Chimica, Universita' di Modena, Via Campi 183, I-41100 Modena, Italy

De Benedetti, P.G. Dipartimento di Chimica, Universita' di Modena, Via Campi 183, I-41100 Modena, Italy

Donnelly, D. Department of Biochemistry and Molecular Biology, University of Leeds, Leeds LS2 9JT, UK

Fanelli, F. Dipartimento di Chimica, Universita' di Modena, Via Campi 183, I-41100 Modena, Italy

Finbow, M.E. Beatson Institute, Garscube Estate, Switchback Road, Bearsden, Glasgow G61 1BD, UK

Findlay, J.B.C. Department of Biochemistry and Molecular Biology, University of Leeds, Leeds LS2 9JT, UK

Harrison, M.A. Department of Biochemistry and Molecular Biology, University of Leeds LS2 9JT, UK

Hurrell, C.R. Department of Biochemistry and Molecular Biology, University of Leeds, Leeds LS2 9JT, UK

IJzerman, A.P. Leiden-Amsterdam Center for Drug Research, Division of Medicinal Chemistry, Gorlaeus Laboratories, P.O. Box 9502, NL-2300 RA, The Netherlands

Jones, P.C. Department of Biochemistry and Molecular Biology, University of Leeds LS2 9JT, UK

Kim, Y.-I. Department of Biochemistry and Molecular Biology, University of Leeds LS2 9JT, UK

Kuipers, W. Department of Medicinal Chemistry, Solvay Duphar B.V., P.O. Box 900, NL-1380 DA Weesp, The Netherlands

Lybrand, T.P. University of Washington, Center for Bioengineering, P.O. Box 351750, Seattle, WA 98195-1750, USA

Menziani, M.C. Dipartimento di Chimica, Universita' di Modena, Via Campi 183, I-41100 Modena, Italy

Milpetz, F. European Molecular Biology Laboratory, Postfach 102209, Meyerhofstrasse 1, D-69012 Heidelberg. Germany

Nemeth, K. Glaxo Institute for Molecular Biology, Plan-les-Quates, CH-1228 Geneva, Switzerland

Oliveira, L. Department of Biophysics, Escola Paulista de Medicina, Sao Pualo 04043-971, Brasil

Paiva, A.C.M. Department of Biophysics, Escola Paulista de Medicina, Sao Paulo 04043-971, Brasil

Persson, B. Department of Medical Biochemistry and Biophysics, Karolinska Institute, S-17199 Stockholm, Sweden

Rippmann, F. E. Merck, Preclinical Pharmaceutical Research, D-64271 Darmstadt, Germany

Sander, C. European Molecular Biology Laboratory, BIOcomputing department, Meyerhofstrasse 1, D-69012 Heidelberg, Germany

Sankararamakrishnan, R. Department of Physiology and Biophysics, Beckman Institute, University of Illinois at Urbana-Champaign, Urbana, IL 61801, USA

Sansom, M.S.P. Laboratory of Molecular Biophysics, University of Oxford, The Rex Richards Building. South Parks Road, Oxford OX1 3QU, UK

Tennant, M. SmithKline Beecham Pharmaceuticals, Computational Chemistry Department, New Frontiers Science Park (North), Third Avenue, Harlow, Essex CM19 5AW, UK

Thomas, P. Department of Biomolecular Structure, Glaxo Research and Development Limited, The Medicines Research Centre, Gunnels Wood Road, Stevenage, Hertfordshire SG1 2NY, UK

Turcatti, G. Glaxo Institute for Molecular Biology, Plan-les-Quates, CH-1228 Geneva, Switzerland

Vriend, G. European Molecular Biology Laboratory, BIOcomputing department, Meyerhofstrasse 1, D-69012 Heidelberg, Germany

Zuurmond, H.M. Leiden-Amsterdam Center for Drug Research, Division of Medicinal Chemistry, Gorlaeus Laboratories, P.O. Box 9502, NL-2300 RA, The Netherlands.

Abbreviations

5-CT	5-carboxamidotryptamine
5-HT	5-hydroxytryptamine
ASP	atomic solvation parameter
B-M	benzophenone-4-maleimide
bR	bacteriorhodopsin
BSA	bovine serum albumin
CFP	channel-forming peptides
DCCD	dicyclohexylcarbodiimide
DHA	(-)-[^{3}H]dihydroalprenolol
DOI	1-(2,5-dimethoxy-4-Br-phenyl)-2-aminopropane
8-OH-DPAT	2-(di-*n*-propylamino)-8-hydroxytetralin
ET	endothelin
F-M	fluorescein-5-maleimide
FTIR	Fourier transform infrared
GA	gramicidin A
GPCR	G protein-coupled receptors
HLA	histocompatibility antigen molecules
IE	interaction energies
LSD	lysergic acid diethylamide
MD	molecular dynamics
nAChR	nicotinic acetylcholine receptor
NBD	7-nitrobenz-2-oxa-1,3-diazol-4-yl
NK	neurokinin
PBS	phosphate-buffered saline
QMD	quenched molecular dynamics
QSAR	quantitative structure–activity relationship
SA/MD	stimulated annealing via restrained molecular dynamics
SAR	structure–activity relationship
SDS–PAGE	sodium dodecyl sulphate–polyacrylamide gel electrophoresis
TCA	trichloracetate
TM	transmembrane

Preface

Integral membrane proteins

The membrane has represented one of the most intractable elements of the biological cell. From the time when it was realized that there was a barrier within the plant cell wall and that this barrier represented the most important and fundamental regulator of the cellular environment, considerable effort has been expended to determine its nature and properties. The lipid fraction was the first to be described but although that was achieved some time ago and interest has waned somewhat since then, there are still fundamental mysteries to explain, not least the real link between function and composition. Knowledge of the protein fraction lagged some way behind, principally due to the intractability of these polypeptides to the methodologies then available. But with the advent/use of such simple molecules as detergents and organic solvents has come the dramatic advances in our understanding of membrane proteins from the primitive days of scaffold-like fragments to the tantalizing mechanisms suggested by the high-resolution 3-dimensional structures of the photosynthetic reaction complex and cytochrome oxidase.

Now that the molecular description of biology is in full flood, it is clearly important to be able to describe the structure and mechanisms of action of integral membrane proteins, both the types that are responsible for the translocation of material and those which mediate the transfer of information. But this urgent need has not so far been fulfilled because yet again, routine methodology is still not available. This has given rise to alternative strategies to gain impressions, albeit crude and unreliable ones, of the structures of integral membrane proteins. The usefulness of such representations in facilitating experimental design and in stimulating the development of new concepts should not be underestimated, however, for there is ample evidence of their predictive potential at least at a low resolution level. During the conference of which these are the proceedings, many of the strategies used to further our structural understanding of integral membrane proteins were described and discussed. Many models were presented and their strengths, weaknesses, contradictions and inconsistencies thoroughly explored.

G protein-coupled receptors

Most attention was devoted to the G protein-coupled receptors (GPCRs), surely now the most avidly pursued and widespread family of proteins in eukaryotic biology. Little did those few groups who wrestled with the mechanism of action, sequence and topography of the visual pigment rhodopsin appreciate the avalanche of structural and functional relatives that lie in the various eukaryotic genomes. The elucidation of the seven transmembrane segments of opsin has given rise to one of the most potent signatures of any protein family. There are now in the database over 700 (and rising) sequences which are recognized by this composite signature. Clearly, this is still only a small proportion of the total if the estimates of the number of olfactory receptors alone are in any way accurate. Already, however, interesting divisions are appearing which in some ways reflect evolutionary distance. The classical sequence-based motifs are incapable of recognizing sub-families such as those for yeast mating factor receptors, the dictyostelium cAMP receptors, the secretin subgroup and receptors for glutamate, Ca, gonadotrophin hormones etc. All appear to be involved in some way with G proteins but perhaps the mechanism of interaction and activation differs.

The assumption, so far unproven, is that the basic framework of these receptors remains more or less intact but considerable functionality is incorporated, often in association with new structural elements or additional domains. Thus, as the ligand gets larger, as with the peptide receptors, the hydrophilic surface regions including the extended N-terminus are recruited in to generate specific binding pockets. This development reaches full expression in the large N-terminal extensions seen in the follicle-stimulating hormone, thyroid-stimulating hormone and luteinizing hormone receptors which have a major role in interacting with the protein ligand. Even larger domains also occur in the glutamate and calcium receptors but here it is harder to give a rational explanation for these regions given the small size of the ligand. Whatever their precise role, these large domains are nevertheless both structurally and functionally intimately associated with the transmembrane segments through which the information flow must pass. It is to be expected that additional elements of structure will also be found at the intracellular face of the receptor. This has already been seen in the cephalopod visual receptors which possess a very unusual region at the C-terminus, one that intriguingly occurs in other, unrelated proteins.

On the basis of limited biophysical evidence, bacteriorhodopsin was the template used for the first tertiary structure representations of GPCRs. This assumption was justified when the low resolution structure of rhodopsin appeared. But it is also clear that the template was only an approximate one for there is variation in the relative position and pitch of some of the transmembrane helices. It is reasonable to assume that rhodopsin is now a more accurate template but, at the same time, it would not be unexpected to find that further differences occur throughout the family and more likely still amongst the more distant and less conserved sub-families.

This conference was designed as a discussion forum to explain and examine the various approaches to modelling integral membrane proteins in general and GPCRs in particular. It revealed wide differences in interpretation particularly of the mutagenesis data. The protein models themselves, despite having different origins and being generated by different methods, apparently had a surprising degree of convergence. The really striking differences were seen in the area of ligand docking where quite different interpretations were put on much the same sets of data. The inevitable conclusion from this was that models whilst stimulating, provocative and reasonably predictive required much firmer structure-based data before one could reliably move from low resolution representations to higher resolution 'structures'.

Finally, one should not overlook the evolving but still unconfirmed concepts in GPCR structure/function relationships. The ternary complex model and the suggestions of a range of conformational intermediates between the activated R* (agonist binding) and inactive R (antagonist binding) modes present new structural challenges. So too do the increasingly frequent suggestions of different binding epitopes for agonists and antagonists, particularly when the former is large and the latter is relatively small. The even more detailed models of GPCRs will have to take on board these subtle concepts and observations. Part of this appreciation must involve a rational description of the activated state, particularly as it applies to the fascinating array of constitutively active mutants.

Transporters/channels

Whilst there is some structural basis for modelling GPCRs, the situation for transporters/channels is much more desperate, at almost all levels of resolution and types of data. The best information so far comes from the structure of bacterial outer membrane porins but it is not at all clear how representative these will be of the majority of channels, since the entire membrane domain consist of an emaculate β–barrel. The suggestions made for the acetylcholine (nicotinic) receptor are intriguing in that a bundle of helices appear to provide the transport route but these may be embedded in a continuous shell of β-sheet. Such data as exist for other transporters/channels emphasize a substantial α-helical content, sometimes incorporating a channel-lining segment of unknown secondary structure. Thus, it looks as though we may have a fascinating mixture of types with much scope for imagination. However, there are both modelling and experimental approaches which might provide useful insights into the structural and functional properties of proteins which mediate material transport. A few examples of these are presented here to illustrate their potential application to model construction.

The meeting was entitled: ‘Membrane Protein Models: Experiment, Theory and Speculation’ (Leeds, UK, March/April 1994, under the auspices of The Molecular Graphics Society). From the excellent presentations that are included in this text, the reader will quickly appreciate the lively and stimulating debates that took place on all three aspects of the structure and function of integral membrane proteins.

J.B.C. Findlay

Chapter 1

Prediction of transmembrane segments in proteins using multiple sequence alignments

Bengt Persson, Frank Milpetz and Patrick Argos

1. Introduction

Membrane proteins are important for several processes and functions in all biological systems. For example, they act as receptors for neurotransmitters or hormones (Savarese and Fraser, 1992; Stephenson, 1991), form a wide variety of ion channels (Barnard, 1992; Miller, 1991), or serve as the respiratory chain (Capaldi, 1991) and transport proteins for different molecules (Griffith *et al.*, 1992; Marger and Saier, 1993; Schloss *et al.*, 1992). Bacterial toxins which form membrane pores (Li, 1992) also belong to this large group of lipid-associated molecules. There are several surface molecules which are anchored to the membrane by one transmembrane segment, for example histocompatibility antigen molecules (HLA), neuraminidases and haemagglutinins (cf. Popot and de Vitry, 1990).

Despite their biological significance tertiary structures have been determined for only a few membrane proteins. Three-dimensional structures at medium to high resolution are available for bacteriorhodopsin (Henderson *et al.*, 1990), photosynthetic reaction centre (Deisenhofer *et al.*, 1985), light-harvesting complexes (Kühlbrandt *et al.*, 1994; McDermott *et al.*, 1995), prostaglandin H_2 synthase-1 (Picot *et al.*, 1994), porin (Weiss and Schulz, 1992) and cytochrome *C* oxidase (Iwata *et al.*, 1995). Other structures are not yet fully resolved, for example photosystem I (Krauss *et al.*, 1993) and nicotinic acetylcholine receptor (Unwin, 1993). Given the scarcity of tertiary structural information, many experimental methods have been applied to determine membrane topology (Jennings, 1989), including analyses of gene fusion proteins and studies of biochemically modified membrane proteins (cf. Traxler *et al.*, 1993).

Membrane proteins constitute a ubiquitous group of structures with members representing several different types of molecular architecture. However, since each traverses the lipid bilayer once or several times, they generally possess hydrophobic sequence segments. Various prediction methods use this characteristic to determine the location of these membrane-spanning regions, albeit with varying degrees of accuracy.

2. Present prediction methods

Theoretical prediction algorithms have been shown to be useful in detecting membrane-spanning segments from the primary structure alone, especially as an aid to designing experiments investigating protein topology. One of the most widely used is that of Kyte and Doolittle (1982), where mean residue hydrophobicity values are calculated for consecutive 19-residue sequence spans. Segments with hydrophobicity above a certain threshold are predicted to be membrane-spanning. A similar approach is adopted by Rao and Argos (1986), who also considered residues that break the transmembrane helices in order to improve reliability of prediction.

Different prediction methods were reviewed and evaluated in 1990 by Degli Esposti *et al.* They examined the correlations amongst the various amino acid hydrophobicity scales used and compared the accuracy of the various prediction approaches. They also calculated a new set of parameters derived from seven different scales (Degli Esposti *et al.*, 1990). A trapezoidal sliding window was used by von Heijne in his hydrophobic analysis of the sequence together with a consideration of positively charged residues interior to the membrane (von Heijne, 1986) particularly with application to the topology of bacterial inner membrane proteins (von Heijne, 1992). These rules were also applied to a number of eukaryotic membrane proteins (Sipos and von Heijne, 1993). Several studies have also been effected regarding helix–helix interactions in membrane proteins (e.g. Lemmon and Engelman, 1992).

Here we present a transmembrane helix prediction algorithm, based upon multiple sequence alignments of related proteins. This method takes advantage of the extended information not found in analysis of a single sequence as is characteristically used by other approaches. The algorithm is described in detail elsewhere (Persson and Argos, 1994). Present primary structural databases are large and expanding at such a rate that homologous sequences are often found. We show that this technique has higher accuracy in predicting transmembrane segments than previous methods based on individual sequences.

3. Residue distributions in membrane proteins

The tertiary structure of only a few membrane proteins are known. However, other types of data exist from various experiments and predictions to deduce the location of transmembrane segments. All sequence segments annotated in the Swissprot database of known primary structures (Bairoch and Boekmann, 1992) as transmembranous were collected. These amounted to 7556 segments from 2319 proteins. The amino acid compositions were investigated for each of 15 positions on the N-terminal and C-terminal side of the centre in the transmembrane segment as indicated in the database. These results were listed with the amino acid residues in the hydrophobicity order of Kyte and Doolittle (1982) (*Table 1*).

The amino acid types were divided into three groups depending on their associated hydrophobicity. The hydrophobic group comprised amino acids Ile, Val, Leu, Phe, Cys, and Met, while the hydrophilic group included Asp, Gln, Glu, Asn, Lys, and Arg. The remaining amino acid residues were ascribed to the intermediate group. A histogram shows the distribution of these three groups at each position of the transmembrane segment (*Figure 1*). There is a clear difference in residue group distribution at position 10 towards the N-terminus and position 11 towards the C-terminus. Between these positions, hydrophobic residues are common and charged residues sparse, while in the flanking regions the opposite is the case. It thus seems likely that the hydrophobic stretch of transmembrane segments is generally 21 amino acids in length. This is compatible with the three-dimensional structure of the reaction centre modelled into a lipid bilayer (Yeates *et al.*, 1987) and corresponds to a membrane-traversing helical structure 30 Å in length along the central axis.

4. Propensity values

For the 2319 proteins in the Swissprot database with a total of 7556 transmembrane segments, 21 residues were assigned to transmembrane spans used for the calculations in this work (i.e. 10 residues were taken on either side of the central residue in the extracted segments which had an odd-numbered length while, for an even-numbered span, 11 were taken from the C-terminal and 10 from the N-terminal sides, respectively).

Amino acid propensity values were calculated with the formula:

$$P^i = \frac{n_{i,\mathrm{seg}} / n_{\mathrm{all,seg}}}{n_{i,\,\mathrm{total}} / n_{\mathrm{all,\,total}}}$$

Here P^i denotes the propensity value of residue type i, and $n_{i,\,\mathrm{seg}}$ and $n_{i,\,\mathrm{total}}$ are the number of ith type residues in the transmembrane segments and in the entire sequences of the 2319 protein data set, respectively. Similarly, $n_{\mathrm{all,\,seg}}$ and $n_{\mathrm{all,\,total}}$ represent the number of all amino acid residues in the transmembrane segments and in the entire membrane proteins, respectively. The P values are effectively the ratio

Table 1. Residue distribution for each position in transmembrace segments of proteins extracted from the Swissprot database

Amino acid	Position N-terminally from centre[a]														
	15	14	13	12	11	10	9	8	7	6	5	4	3	2	1
Ile	346	436	518	459	468	887	877	921	898	815	809	1024	772	893	924
Val	280	365	425	433	550	742	807	775	919	719	887	883	991	909	930
Leu	545	553	765	836	862	1207	1080	1306	1162	1232	1239	1318	1345	1339	1191
Phe	317	346	352	294	395	619	548	672	611	691	682	676	654	661	763
Cys	58	110	156	94	79	77	88	137	159	173	147	200	236	235	236
Met	106	176	202	201	227	306	298	251	230	326	286	208	322	204	179
Ala	425	449	408	482	631	524	695	723	841	612	759	834	885	1001	813
Gly	478	406	514	559	378	501	546	550	560	512	566	623	628	538	645
Thr	358	468	392	403	429	310	389	395	361	488	347	317	321	307	415
Ser	502	533	433	417	529	334	378	430	390	438	513	392	381	430	448
Trp	90	208	236	205	164	200	263	169	185	157	209	205	147	111	159
Tyr	190	243	271	280	316	328	296	183	229	205	265	160	139	151	104
Phe	320	376	278	452	264	189	295	177	128	175	111	123	171	165	123
His	148	207	156	165	156	147	101	203	237	117	98	34	50	22	68
Asp	181	226	262	275	197	160	102	63	44	104	92	95	54	68	56
Gln	339	236	376	264	345	65	61	40	68	92	51	34	25	35	25
Glu	325	388	371	334	399	140	166	75	114	144	95	84	107	135	109
Asn	217	238	258	236	166	125	110	82	77	183	77	56	41	54	57
Lys	319	367	380	458	329	188	79	85	39	40	27	26	27	11	50
Arg	435	476	474	438	416	278	148	90	75	104	67	34	31	58	32

Amino acid	Position C-terminally from centre[a]														
	1	2	3	4	5	6	7	8	9	10	11	12	13	14	15
Ile	761	837	1009	837	856	790	807	904	866	711	744	497	437	412	384
Val	922	992	981	855	934	887	933	838	815	774	654	474	452	377	397
Leu	1219	1508	1243	1245	1230	1371	1249	1055	1249	1199	843	678	797	600	545
Phe	677	676	649	584	675	619	602	563	670	654	494	387	270	439	260
Cys	245	157	198	192	145	157	195	176	160	180	164	204	127	107	129
Met	219	196	184	267	222	245	251	327	307	225	245	168	212	152	209
Ala	839	802	615	669	820	683	663	724	759	700	613	458	482	441	380
Gly	523	483	638	703	571	542	497	509	595	418	467	563	373	460	435
Thr	420	342	515	407	471	491	352	477	345	356	323	397	355	405	388
Ser	561	514	432	483	390	395	395	456	341	387	359	463	566	460	516
Trp	185	114	138	144	81	160	165	143	146	194	239	218	182	97	125
Tyr	155	135	117	142	229	204	280	314	387	508	448	350	329	427	300
Phe	246	217	235	217	219	284	337	178	113	159	204	269	269	449	262
His	51	20	47	56	65	60	92	128	111	132	171	181	185	162	143
Asp	26	41	26	32	25	36	33	45	40	57	163	253	217	263	252
Gln	68	85	82	92	33	44	65	124	61	129	244	344	241	389	351
Glu	98	72	101	173	190	210	189	127	123	170	217	292	391	292	291
Asn	58	54	60	167	102	60	102	111	96	83	165	226	239	270	260
Lys	17	26	22	30	46	52	40	52	78	151	255	390	442	414	410
Arg	37	55	35	32	23	37	80	76	65	140	304	488	704	554	557

[a]The number of occurrences for each of the 20 different amino acid types at positions 1–15 N-terminally (page 4) and C-terminally (page 5) counted from the central position in the transmembrane sequence segments is shown.

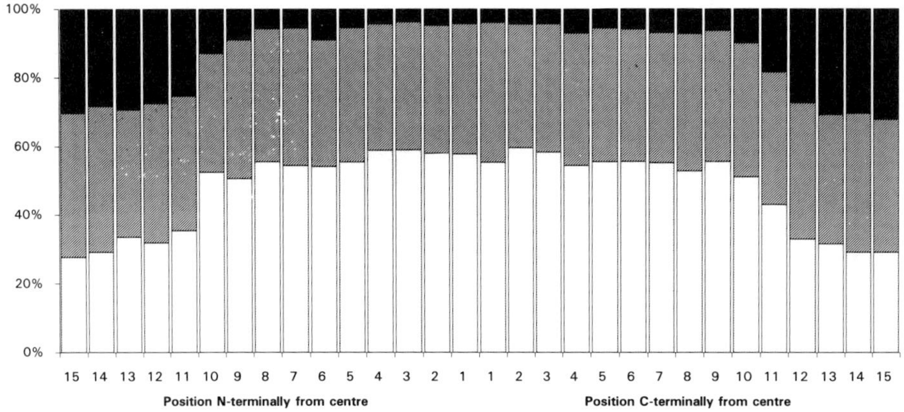

Figure 1. Positional distribution of amino acid residue types in transmembrane segments. The histogram shows the amino acid compositions for 15 N- and C-terminal positions relative to the centre of putative transmembrane segments listed in the feature tables of Swissprot. The data from *Table 1* are used and expressed as the percentage contribution of each amino acid type. For clarity, the residues are divided into three different groups dependent upon their hydrophilicity. Dark segments show hydrophilic residues (Arg, Lys, Asn, Glu, Gln, and Asp); hatched segments give intermediate residues (His, Phe, Tyr, Trp, Ser, Thr, Gly, and Ala); light stippled segments represent hydrophobic residues (Met, Cys, Phe, Leu, Val, and Ile). The compositions of positions 11–15 at the N-terminal side and 12–15 at the C-terminal side differ significantly from the others, especially for the most hydrophobic and charged/hydrophilic residues, suggesting that transmembrane spans generally consist of a hydrophobic region 21 residues in length.

of the fraction in which an amino acid type appears in membrane spans to that in the entire protein sequences. Values above 1.00 indicate a preference to be in the lipid-associated structure while below 1.00 an avoidance is implied.

Two sets of P values were calculated: one for the 21-residue hydrophobic region of transmembrane segments (denoted $P_m^{i)}$ and one for the flanking four-residue extensions at both termini (P_e^i), corresponding to those amino acids likely to interact with the phosphate groups of the membrane lipids (*Table 2*). The amino acid residues with high P_m^i values are predominantly hydrophobic, as expected. The P_e^i values show a clear tendency towards charged and polar residues. However, the basic amino acids Arg and Lys have a higher tendency to be at the termini of transmembrane segments than the acidic Asp and Glu, which are not likely partners for phosphate groups. The high preponderance of Trp and Tyr is in concordance with findings of Sipos and von Heijne (1993).

Table 2. Propensity values for amino acid residue types in the hydrophobic central (P_m) and terminal (P_e) regions of transmembrane helical spans

Amino acid	P_m	P_e
Ala	1.409	0.888
Arg	0.240	1.519
Asn	0.434	1.103
Asp	0.192	0.740
Cys	1.069	0.842
Gln	0.344	0.889
Glu	0.175	0.804
Gly	1.058	0.920
His	0.588	1.117
Ile	1.990	1.103
Leu	1.702	0.998
Lys	0.181	1.178
Met	1.501	1.172
Phe	1.966	1.102
Pro	0.519	0.881
Ser	0.774	0.920
Thr	0.828	0.881
Trp	1.314	1.450
Tyr	0.979	1.314
Val	1.694	0.870

5. Prediction algorithm

(1) First, a multiple sequence alignment is created. This can be achieved using any suitable program. In the development of the algorithm, we used PILEUP from the GCG package (Devereux *et al.*, 1984). In order to avoid bias towards closely related sequences, each sequence i in the alignment is assigned a weight (w_i) based upon its dissimilarity with all other aligned sequences according to Vingron and Argos (1989). The number of residue identities in each pairwise comparison provides the similarity criterion.

(2) For each sequence i and for each site in the sequence corresponding to position j in the multiple alignment, $P_m{}^{i,j}$ and $P_e{}^{i,j}$ values are calculated as the average P_m and P_e preference values over windows of 15 and four residues, respectively. These values are assigned to each sequence site where the alignment position j corresponds to the central window position. This process is repeated for all possible 15- and four-residue segments in each sequence.

(3) For each position j in the alignment, $\langle P_m \rangle^j$ and $\langle P_e \rangle^j$ are calculated as the weighted mean over the ith sequence sites at alignment position j with associated $P_m^{i,j}$ and $P_e^{i,j}$ values; that is:

$$\langle P_m \rangle^j = \sum_{i=1}^{N} w_i P_m^{ij} \Big/ \sum_{i=1}^{N} w_i$$

(4) Cores of possible transmembrane regions are identified from segments of eight or more consecutive alignment positions with $\langle P_m \rangle^j$ values above 1.23. A type of smoothing is performed by also allowing those spans lacking one or at most two positions with $\langle P_m \rangle$ values below the limit to be included, providing one of the neighbouring $\langle P_m \rangle$ values was greater than 1.23. If an alignment position contains exclusively any one of the charged residue types Asp, Glu, Arg, or Lys, $\langle P_m \rangle$ greater than 1.18 was sufficient to allow incorporation into the eight-position consecutive span. This procedure was undertaken in order to adjust for functional groups in the transmembrane segments (cf. Henderson *et al.*, 1990; Savarese and Fraser, 1992).

(5) The alignment segments of eight positions thus detected are elongated N- and C-terminally successively by one position in the direction of the highest flanking $\langle P_m \rangle$ value, such that $\langle P_m \rangle$ is greater than 1.17 for each extension site and the total segment length does not exceed 21 residues. Segments with overlapping alignment positions are fused.

(6) $\langle P_e \rangle^j$ values are checked within five positions on either side of the termini obtained by rule (5). Peak values above 1.08 are then used to determine the start and end points of the transmembrane region providing the $\langle P_e \rangle$ peaks were 20–33 positions apart and the extension conditions of rule (5) are met. If a terminal pair of $\langle P_e \rangle$ peaks are both over 1.15, then the extension $\langle P_e \rangle$ threshold is lowered to 1.08 from 1.17 (rule (5)). If the above conditions are not met, then only $\langle P_m \rangle$ values in rule (7) below are used to delineate the extent of the transmembrane region.

(7) Segments are split if they are sufficiently long to contain multiple transmembrane helices of a minimum length of 21 residues each and with at least seven positions separating each potential helical region. Segments directly found or resulting from previous procedures that are shorter than 15 positions are eliminated; those longer than 29 positions that cannot be split are shortened to the 29 consecutive positions containing the largest mean $\langle P_m \rangle$ values. In both cases, the lengths are averaged over the different sequences, excluding gaps.

(8) The predicted region of the multiple alignment is then applied directly to equivalent segments in each of the single sequences.

6. Optimization of the method

The thresholds used in this method were obtained by systematic testing of different parametric values. Window sizes of 7–21 and 2–5 in steps (2) and (1) above were utilized in the calculations of $P_m^{i,j}$ and $P_e^{i,j}$, respectively. Limits for $\langle P_m \rangle$ in steps (4) and (5) above were initially tested in the range 1.00–1.30 with a step size of 0.02, and then further scrutinized around the optimal values with a step of 0.01. Similarly, various $\langle P_e \rangle$ thresholds were tested in the range 1.04–1.12 with a step size of 0.01, with the $\langle P_m \rangle$ limits set to the optimal values previously determined. The final length criterion of the transmembrane segments was tested with miniumum values of 15, 16, and 17 and maximum values of 29, 31, and 33, where the other previously determined optimal thresholds were taken. For step (7), values of 6, 7, and 8 positions for separating the potential helices were considered. The values presented above were those found to be optimal, such that the number of incorrectly predicted transmembrane segments was smallest for the test set of 28 protein families (*Table 3*).

7. Computer program

A computer program, TMAP, has been written in C to calculate the transmembrane regions in a multiple sequence alignment according to the new algorithm. The results of this program are presented as positional numbers both for the entire alignment and for each of the included single amino acid sequences. An example output is given in *Figure 2*. In addition, a PostScript plot of the $\langle P_m \rangle$ and $\langle P_e \rangle$ values against the alignment position is produced. An example for adrenergic receptor sequences is shown in *Figures 2* and *3*. There are seven obvious regions with high membrane propensity values ($(\langle P_m \rangle)$); these segments correspond to the seven transmembrane segments (*Figure 2*). The predictions are shown by the top horizontal bars, where the thick parts delineate the central regions and the thin parts, the four-position flanks (*Figure 3*).

The source code of the program, written in C, is available from EMBL via anonymous ftp (ftp.embl-heidelberg.de), Gopher or WorldWideWeb also referred to as Mosaic (URL is http://www.embl-heidelberg.de/tmap-info.html). The code can be compiled on most types of computers, provided they have sufficient memory. If this is not the case, the dimensions given in the code must be modified.

8. Electronic mail

There is also an electronic mail (e-mail) service for predictions of transmembrane segments. After submission of the multiple sequence alignment in the GCG format (.msf file) or in the PIR format to the address 'tmap@embl-heidelberg.de', the TMAP prediction program is run, and subsequently the result files returned via e-mail. Instructions are available by sending the message 'HELP' to the above address.

A single sequence may also be sent by e-mail in the GCG or PIR format. In this case, a BLITZ search of the entire Swissprot protein sequence database for homologous proteins is performed using the procedure of MPSEARCH (Collins

Table 3. Test of 28 protein families used in the evaluation of the prediction algorithm[a]

Entry	Code	Family name / Aligned proteins (and total number)	Residue identity (%) in pairwise comparisons
1	ad	Adrenergic receptors	26.5–96.9
		A2ab_Human, A2ab_Rat, A2aa_Human, A2aa_Pig, A2aa_Rat, A2ac_Human, A2ac_Rat, A1aa_Human, A1aa_Rat, A1ab_Mesau, A1ab_Rat, A1ab_Canfa, A1ac_Bovin, B3ar_Mouse, B3ar_Rat, B3ar_Human, B2ar_Mouse, B2ar_Rat, B2ar_Mesau, B2ar_Human, B1ar_Human, B1ar_Rat, B1ar_Melga (23)	
2	am	Muscarinic acetylcholine receptors	35.4–99.1
		Acm2_Human, Acm2_Pig, Acm2_Rat, Acm4_Human, Acm4_Rat, Acm4_Chick, Acm3_Human, Acm3_Pig, Acm3_Rat, Acm1_Human, Acm1_Pig, Acm1_Rat, Acm1_Mouse, Acm5_Human, Acm5_Rat, Acm1_Drome (16)	
3	an	Nicotinic acetylcholine receptors	28.5–99.0
		Achb_Mouse, Achb_Rat, Achb_Bovin, Achb_Human, Achb_Torca, Achd_Muse, Achd_Rat, Achd_Bovin, Achd_Chick, Achd_Xenla, Achd_Torca, Ache_Mouse, Ache_Rat, Ache_Bovin, Achg_Mouse, Achg_Rat, Achg_Bovin, Achg_Human, Achg_Chick, Achg_Xenla, Achg_Torca, Ach1_Schgr, Ach2_Drome, Ach1_Drome, Ach4_Drme, Ach5_Chick, Ach5_Rat, Acho_Carau, Achp_Carau, Acho_Rat, Achp_Chick, Achp_Rat, Achn_Human, Achn_Rat, Achn_Chick, Achn_Caru, Ach1_Xenla, Ach2_Xenla, Acha_Torca, Acha_Torma, Acha_Mouse, Acha_Rat, Acha_Bovin, Acha_Human, Acha_Chick, Ach3_Chick, Ach3_Rat, Ach3_Carau, Ach2_Chick, Ach2_Rat, Ach4_Chick, Ach4_Rat, Ach3_Drome, Ach7_Chick (54)	
4	b3	Band 3 proteins	50.1–98.3
		B3at_Mouse, B3at_Rat, B3at_Human, B3at_Chick, B3a3_Mouse, B3a3_Rat, B3lp_Mouse, B3lp_Rat, B3lp_Human (9)	
5	ba	Bacteriorhodopsins	20.8–57.8
		Baca_Hals1, Bacr_Halha, Bach_Halsp, Bach_Natph, Bacs_Halha (5)	
6	cb	Chemoreceptors	39.2–78.3
		Mvp2_Ecoli, Mcp2_Salty, Mcp1_Ecoli, Mcps_Entae, MCp4_Ecoli, Mcpd_Entae, Mcp3_Ecoli (7)	

7	cn	Connexins Cx45_Canfa, Cx45_Mouse, Cs45_Chick, Cx26_Human, Cx26_Rat, Cx32_Human, Cs32_Rat, Cx32_Mouse, Cx30_Xenla, Cx31_Mouse, Cx31_Rat, Cx30_Rat, Cx40_Rat, Cx42_Chick, Cx50_Mouse, Cx43_Mouse, Cx43_Rat, Cx43_Human, Cx43_Bovin, Cx43_Chick, Cx43_Xenla, Cx33_Rat, Cx37_Mouse, Cx38_Xenla (24)	28.7–99.7
8	co	Cytochrome oxidases Cox2_Ascsu, Cox2_Caeel, Cox2_Gorgo, Cox2-Human, Cox2_Panpa, Cox2_Hylsy, Cox2_Cerae, Cox2_Macfa, Cox2_Mouse, Cox2_Rat, Cox2_Micpe, Cox2_Balph, Cox2-Bovin, Cox2_Phovi, Cox2_Cypca, Cox2_Xenla, Cox2_Chick, Cox2_Parli, Cox2_Pisoc, Cox2_Drome, Cox2_Droya, Cox2-Locmi, Cox2_Apime, Cox2_Maize, Cox2_Orysa, Cox2_Wheat, Cox2_Dauca, Cox2_Hansa, Cox2_Klula, Cox2_Pneca, Cox2-Podan, (31)	30.5–98.7
9	da	Dopamine receptors Dadr_Human, Dadr_Rat., D5dr_Human, Dbdr_Rat, D2dr_Bovin, D2dr_Rat, D2dr_Human, D2dr_Xenla, D3dr_Rat, D4dr_Human (10)	19.9–96.6
10	gb	GABA (γ-amino butyric acid) receptors Gac3_Mouse,Gac3_Rat, Gac2_Mouse, Gac2_Rat, Gac2-Bovin, Gac2_Human, Gac2_Chick, Gac1_Rat, Gaa3_Mouse, Gaa3_Rat, Gaa 3_Bovin, Gaa2_Mouse. Gaa2_Rat, Gaa2_Bovin, Gaa1_Bovin, Gaa1_Human, Gaa1_Rat, Gaa1_Chick, Gaa5_Rat, Gaa4_Bovin, Gaa4_Rat, Gaa6_Mouse, Gar1_Human, Gar2_Human, Gad_Mouse, Gad_Rat, Gab1_Human, Gab1_Rat,Gab1_Bovin, Gab3_Human, Gab3_Rat, Gab3_Chick, Gab2_Rat, Gab4_Chick, Gab_Lymst, Gab_Drome (35)	29.0–99.8
11	gp	Glycophorins Glpb_Human, Glpe_Human, Glpa_Human, Glp-Mouse, Glp_Pig (5)	18.0–70.8
12	gt	Glucose trasporters Gtr2_Mouse, Gtr2_Rat, Gtr2_Human, Gtr4_Human, Gtr4_Rat Gtr4_Mouse, Gtr1_Mouse, Gtr1_Rat, Gtr1_Bovin, Gtr1_Pig, Gtr1_Human, Gtr1_Rabbit, Gtr3_Chick, Gtr3_Human, Gtr5_Human, Glcp_Syny3, Stp1_Arath, Hxt2_Yeast, Rag1_Klula, Snf3_Yeast (20)	20.8–98.2
13	ha	Haemagglutinins Hema_Pi3h4, Hema_Pi3ha, Hema_Pi3hu, Hema_Pi3hv, Hema_Pi3hw, Hema_Pi3ht, Hema_Pi3hx, Hema_Pi3b, Hema_Send5, Hema_Sendj, Hema_Send2, Hema_Sendf, Hema_Sendh, Hema_Pi1hw, Hema_Ndvb, Hema_Ndvtg, Hema_Ndvh, Hema_Ndvd, Hema_Ndvu, Hema_Ndvq, Hema_Ndvi, Hema_Ndvm, Hema_Ndva, Hema_Mumpm, Hema_Mumpr, Hema_Mumps, Hema_Sv5cm, Hema_Sv5cp, Hema_Sv5, Hema_Sv51n, Hema_Pi2h, Hema_Pi2ht, Hema_Sv41, Hema_Pi4ha (34)	19.9–99.8

Table 3. Continued

Entry	Code	Family name Aligned proteins (and total number)	Residue identity (%) in pairwise comparisons
14	m1	MHC (Major histocompatibility antigen) I Ha1t_Mouse, Ha1u_Mouse. Ha1a_Bovin, Ha1b_Bovin, Ha1a_Rabbit, Ha1b_Rabbit, Ha12_Ponpy, Ha1z_Human, Ha1x_Human, Ha1y_Human, Ha1d_Pantr, Ha1n_Pantr, Ha1m_Human, Ha1p_Human, Ha1w_Human, Ha1o_Human, Ha1n_Human, Ha1q_Human, Ha1u_Human, Ha1t_Human, Ha11_Human, Ha1s_Human, Ha1m_Pantr, Ha11_Human, Ha13_Human, Ha1a_Human, Ha1d_Human, Ha1c_Human, Ha1j_Human, Ha1c_Pantr, Ha1a_Pantr, Ha1b_Pantr, Ha10_Human, Ha11_Ponpy, Ha1g_Human, Ha1k_Human, Ha1v_Human, Ha16_Human, Ha1e_Pantr, Ha19_Canfa, Ha11_Rat, Ha12_Rat, Ha1q_Mouse, Ha1w_Mouse, Ha1k_Mouse, Ha1b_Mouse, Ha1d_Mouse, Ha11_Mouse, Ha13-Mouse, Ha14_Mouse, Ha12_Mouse, Ha17_Mouse, Ha18_Mouse, Ha10_Mouse, Ha15_Mouse, Ha1f_Chick (56)	34.2–99.7
15	m2	MHC (major histocompatibility antigen) II Rng6_Human, Rng6_Mouse, Ha21_Mouse, Ha22_Mouse, Ha23_Mouse, Ha2r_Human, Ha2k_Mouse, Ha2u_Mouse, Ha2b_Mouse, Ha2f_Mouse, Ha2j_Mouse, Ha2r_Mouse, Ha2s_Mouse, Ha2q_Mouse, Ha2d_Mouse, Ha2b_Rat, Ha2c_Pig, Ha2d_Pig, Ha21_Human, Ha24_Human, Ha23_Human, Ha27_Human, Ha22_Human, Ha25_Human, Ha26_Human, Ha2q_Human, Ha2z_Human, Rng7_Human (28)	15.0–99.6
16	m3	MHC (major histocompatibility antigen) II Hb2o_Human, Hb2o_Pantr, Hb2q_Human, Hb2t_Human, Hb2p_Human, Hb2s_Human, Hb21_Spaeh, Hb2p_Rabbit, Hb24_Mouse, Ha2b_Mouse, Hb2i_Mouse, Hb23_Mouse, Hb21_Mouse, Hb22_Mouse, Hb2j_Mouse, Hb2d_Rat, Hb2d_Human, Hb2h_Human, Hb2e_Human, Hb2f_Human, Hb2c_Human, Hb2b_Human, Hb2i_Human, Hb2a_Human, Hb2j_Human, Hb2k_Human, Hb2g_Human, Hb2d_Canfa, Hb2d_Mouse, Hb2q_Mouse, Hb2a_Mouse, Hb2f_Mouse, Hb2s_Mouse, Hb2k_Mouse, Hb2u_Mouse, Hb2a_Rat, Hb2c_Pig, Hb2d_Pig, Hb23_Human, Hb24_Human, Hb21_Human, Hb22_Human, Hb2x_Human, Hb21_Chick (42)	44.2–99.2
17	mi	MIP (major intrinsic proteins) Mip_Bovin, Mip_Rat, Tipa_Arath, Tipa_Phavu, Tipg_Arath, No26_Soybn, Bib_Drome (7)	22.1–93.9

18	na	Neuraminidases	17.4–99.1
		Nram_Iada1, Nram_Iatkr, Nram_Iapue, Nram_Iawil, Nram_Iapar, Nram_Iacao, Nram_Iauss, Nram_Iachi, Nram_Iame1, Nram_Iabda, Nram_Iahk1, Nram_Iash2, Nram_Iackq, Nram_Iackr, Nram_Iaudo, Nram_Iavi7, Nram_Iaban, Nram_Iant6, Nram_Iatok, Nram_Iari5, Nram_Iame2, Nram_Iatra, Nram_Iawhm, Nram_Iadge, Nram_Iafpw, Nram_Iahco, Nram_Iatkp (26) Nram_Inbln, Nram_Inbvi, Nram_Inbbe, Nram_Inbmf, Nram_Inbor, Nram_Inbsi, Nram_Inbhk, Nram_Inbus, Nram_Inbmd, Nram_Inble (10)	
20	op	Opsins	18.3–96.3
		Opsg_Human, Opsr_Human, Opsg_Astfa, Opsh_Astfa, Opsr_Astfa, Opsr_Chick, Opsb_Human, Opsv_Chick, Opsd_Bovin, Opsd_Sheep, Opsd_Crigr, Opsd_Mouse, Opsd_Human, Opsd_Chick, Opsd_Lamja, Opsg_Chick, Opsb_Chick, Ops3_Drome, Ops3_Drops, Ops4_Drome, Ops4_Drovi, Opsd_Lolfo, Opsd_Octdo, Ops2_Drome, Ops2_Drops, Ops1_Drome, Ops1_Drops, Ops1_Calvi (28)	
21	ph	Photosystem II P680 chlorophyll A apoproteins	69.4–99.6
		Psbb_Anasp, Psbb_Syny3, Psbb_Proho, Psbb_Secce, Psbb_Wheat, Psbb_Horvu, Psbb_Maize, Psbb_Orysa, Psbb_Oenho, Psbb_Tobac, Psbb_Spiol, Psbb_Marpo, Psbb_Euggr (13)	
22	ps	Photosystem Q(B), II D2 proteins	24.4–99.7
		Psbd_Orysa, Psbd_Secce, Psbd_Horvu, Psbd_Pea, Psbd_Tobac, Psbd_Spiol, Psbd_Marpo, Psbd_Chlre, Psbd_Synp7, Psbd_Syny3, Psbd_Synp2, Psbd_Cyaca, Psba_Anaaz, Psba_Fredi, Psb2_Syny3, Psba_Syny4, Psb2_Synp7, Psb1_Synp7, Psba_Proho, Psba_Horvu, Psba_Secce, Psba_Orysa, Psba_Amahy, Psba_Spiol, Psba_Tobac, Psba_Pethy, Psba_Brana, Psba_Soybn, Psba_Sinal, Psba_Medsa, Psba_Pea, Psba_Spiog, Psba_Marpo, Psba_Vicfa, Psba_Pinco, Psba_Chlmo, Psba_Chlre, Psba_Cyapa, Psba_Antsp, Psba_Euggr, Psba_Cyaca, Psba_Ectsi, Psb1_Syny3 (43)	
23	rc	Photosynthetic reaction centre	15.5–99.4
		Psbd_Secce, Psbd_Spiol, Psbd_Synp7, Psbd_Syny3, Psbd_Synp2, Psba_Soybn, Psba_Spiol, Psba_Sinal, Psba_Secce, Psba_Spiog, Psba_Syny4, Rcem_Rhoru, Rcem_Rhovi, Rcem_Rhoca, Rcem_Rhosh, Rcem_Erysp, Rcem_Chlau, Rcel_Rhoca, Rcel_Rhosh, Rcel_Erysp, Rcel_Rhoru, Rcel_Rhovi, Rcel_Chlau (23)	
24	t1	Sodium/potassium-transporting ATPase β-chains	26.5–96.9
		Atnb_Mouse, Atnb_Rat, Atnb_Human, Atnb_Canfa, Atnb_Sheep, Atnb_Pig, Atnb_Chick, Atnb_Torca, Atnc_Mouse, Atnc_Rat, Atnc_Human, Atnc_Xenla, Atnb_Artsa (13)	

Table 3. Continued

Entry	Code	Family name Aligned proteins (and total number)	Residue identity (%) in pairwise comparisons
25	t2	ATP synthase A chains Atpi_Maize, Atpi_Orysa, Atpi_Spiol, Atpi_Tobac, Atpi_Pea, Atpi_Marpo, Atp6_Syny3 (7)	65.2–99.2
26	t3	Potassium-transporting ATPase α-chains Atha_Human, Atha_Rabit, Atha_Pig, Atha_Rat, Ath1_Schpo, Ath2_Schpo, Ath1_Yeast, Ath2_Yeast, Athp_Zygro, Ath1_Canal, Athp_Neucr, Atxa_Leido, Atxb_Leido, Ath1_Lyces, Ath2_Lyces, Ath1_Arath, Ath2_Arath, Ath3_Arath (18)	20.2–98.6
27	t4	ATP synthase lipid-binding proteins Atph_Pea, Atph_Spiol, Atph_Marpo, Atph_Tobac, Atph_Soybn, Atpl_Synp6, Atpl_Syny3 (7)	79.0–98.8
28	t5	Calcium-transporting ATPases Atcq_Human, Atcq_Pig, Atcp_Rat, Atcq_Rat, Atcr_Human, Atce_Human, Atce_Pig, Atcd_Pig, Atcd_Rat, Atcd_Human, Atcd_Rabit, Atce_Rat, Atce_Rabit, Atca_Rabit, Atcb_Rabit, Atcb_Chick, Atcf_Rat, Atcb_Drome, Atn2_Chick, Atn2_Rat, Atn3_Human, Atn3_Rat, Atn3_Chick, Atn1_Pig, Atn3_Pig, Atn1_Human, Atn1_Sheep, Atn1_Horse, Atn1_Rat, Atn1_Chick, Atna_Catco, Atna_Torca, Atna_Artsf, Atna_Drome, Atna_Artsa (35)	26.7–99.9

[a]In calculations of residue identities from aligned sequence pairs, gaps are excluded. The values given are the maximum and minimum identity percentages over all pairwise sequence comparisons within the family. Each sequence is referred to by its identification name given in the Swissprot database (Bairoch and Boekmann, 1992).

```
RESULTS from program TMAP, edition 46'
Numbers give: a) number of transmembrane segment
              b) start of TM segment (alignment position / residue number)
              c) end of TM segment (alignment position / residue number)
              d) length of TM segment within parentheses

PREDICTED TRANSMEMBRANE SEGMENTS FOR ALIGNMENT a1.msf
   TM 1:  89 -117 (28.4)
   TM 2: 128 -156 (28.3)
   TM 3: 160 -184 (25.0)
   TM 4: 205 -233 (28.4)
   TM 5: 256 -283 (28.0)
   TM 6: 451 -474 (24.0)
   TM 7: 489 -508 (20.0)
PREDICTED TRANSMEMBRANE SEGMENTS FOR PROTEIN A2ab_Human
   TM 1:  11 -  39 (29)
   TM 2:  50 -  78 (29)
   TM 3:  82 -106 (25)
   TM 4: 127 -152 (26)
   TM 5: 171 -198 (28)
   TM 6: 365 -388 (24)
   TM 7: 403 -422 (20)
PREDICTED TRANSMEMBRANE SEGMENTS FOR PROTEIN A2ab_Rat
   TM 1:  16 -  44 (29)
   TM 2:  55 -  83 (29)
   TM 3:  87 -111 (25)
   TM 4: 132 -157 (26)
   TM 5: 176 -203 (28)
   TM 6: 368 -391 (24)
   TM 7: 406 -425 (20)
PREDICTED TRANSMEMBRANE SEGMENTS FOR PROTEIN A2aa_Human
   TM 1:  32 -  60 (29)
   TM 2:  71 -  99 (29)
   TM 3: 103 -127 (25)
   TM 4: 148 -176 (29)
   TM 5: 195 -222 (28)
   TM 6: 368 -391 (24)
   TM 7: 403 -422 (20)
```

Figure 2. Example output from program TMAP. The results for the family of adrenergic receptors (family code 'ad', cf. *Table 1*) are shown. The first prediction gives the positional numbers referring to the alignment. The subsequent predictions are given by the corresponding residue numbering scheme for each of the aligned sequences.

and Coulson, 1990) which relies on the Needleman and Wunsch (1970) alignment procedure and PAM (Dayhoff *et al.*, 1983) or BLOSUM (Henikoff and Henikoff, 1992) residue exchange weight scores. The related sequences found are then aligned with the submitted sequence by the technique of CLUSTAL W (Thompson *et al.*, 1994). The resulting prediction and plot are returned to the submitter as usual.

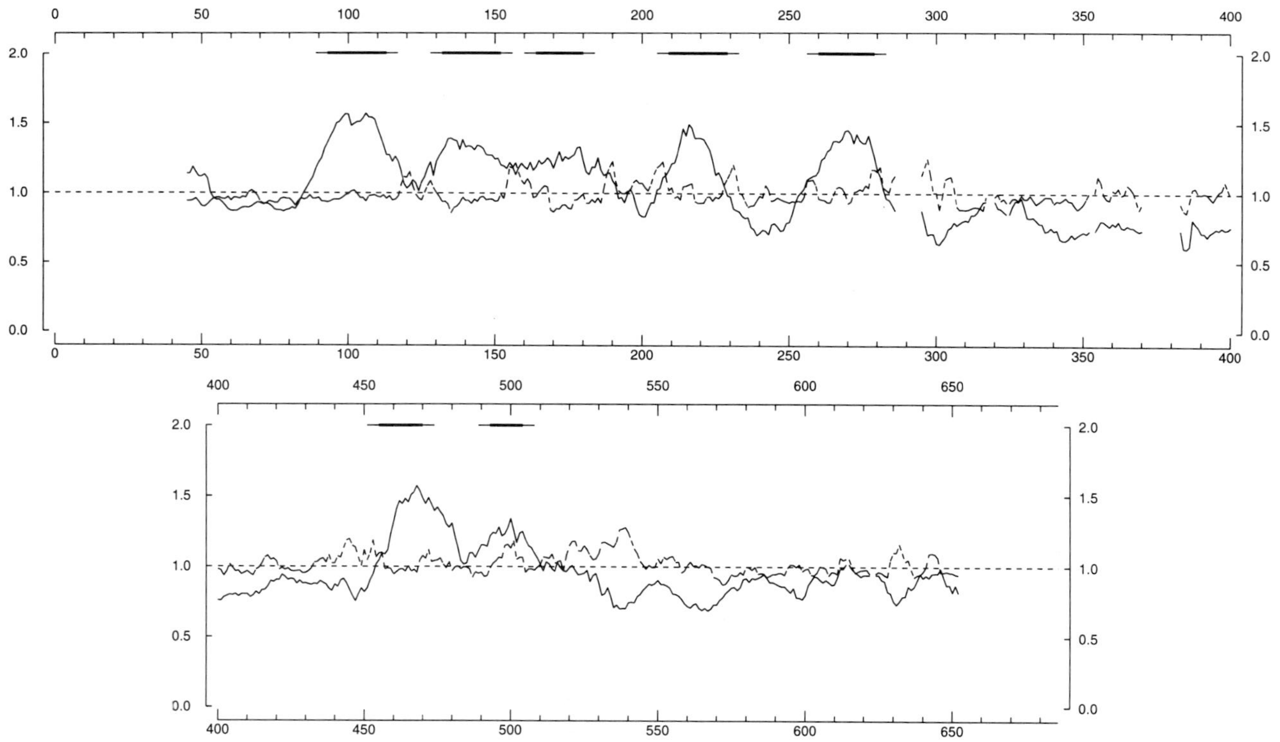

Figure 3. Example plot from program TMAP. The example plot corresponds to the predictions in *Figure 2* for the 'ad' family. The continuous curve gives the mean propensity values, $\langle P_m \rangle^j$ (continuous line) and $\langle P_e \rangle^j$ (broken line) for each alignment position j. The seven predicted segments are indicated by the horizontal bars, where the thick parts correspond to the central, hydrophobic portion and the thin parts to the flanking residues. The regions where no curve is given results from positions in the multiple alignments with amino acid residues only in single sequences and gaps in the other structures; such regions are not predicted.

The single- and multiple-sequence submissions can also be performed through WorldWideWeb (http://www.embl-heidelberg.de/tmap/tmap-info.html). Parametric options used in executing BLITZ (e.g. gap penalty values in amino acid exchange weight matrices) are available, as is also the case in the e-mail queries (see HELP). If none are given, default values are selected automatically and the TMAP program will execute. The prediction results from WorldWideWeb submissions will be sent by normal e-mail.

9. Evaluation of the new algorithm

The new prediction algorithm was evaluated using a test set of 28 different families of membrane proteins, in total having 126 transmembrane segments, extracted from the Swissprot database (release 25, March 1993; Bairoch and Boekmann, 1992) with the aid of the computer program SRS (Sequence Retrieval Software; Etzold and Argos, 1993a,b) and by searching the Prosite pattern database (release 10.0, December 1992; Bairoch, 1992) and Swissprot key words. The families and the proteins included are listed in *Table 3*. The data set contained 633 sequences, representing several folding and functional types with a mixed spread in pairwise sequence identity fraction.

Multiple alignments were created using the automated routine PILEUP (Devereux *et al.*, 1984; Genetics Computer Group, 1991, Wisconsin, USA) with penalties of 6.0 for gap initiation and 0.2 for each gap extension and with alignment scores based upon the Dayhoff PAM 250 amino acid substitution matrix (Dayhoff *et al.*, 1983). Pairwise residue identity amongst all family sequence pairs was typically above 20% and each family had five or more members (*Table 3*).

The predictions were compared with the 'standard-of-truth' as given in the Swissprot database together with known three-dimensional structures. The results are listed in *Table 4*. Since the segment limits given in Swissprot do not exactly correspond to equivalent multiple alignment positions in different but related proteins, the mean borders were used as the expected termini of transmembrane helices. Termini for bacteriorhodopsin were taken from Henderson *et al.* (1990), photosynthetic reaction centre from Yeates *et al.* (1987), and chloroplast atpI from Hudson *et al.* (1987).

The new method showed five mispredictions (i.e. either a segment is not predicted or falsely predicted). This corresponds to a transmembrane segment detection rate of nearly 96%. The termini of the predicted segments differ on average by 3.6 residues from the designations in the Swissprot database.

To test for the bias of including the same proteins in the calculation of the residue preference values as those used in the predictions, a 'jacknife test' was performed

Table 4. Results of predictions for the test set[a]

Protein family	Reference sequence	Transmembrane segment				
		No.	Position	Length	Deviations	
ad	A2aa_Human	1	32–60	28.4	−2	+2
		2	71–99	28.3	0	+3
		3	103–127	25.0	−4	−2
		4	148–176	28.4	−3	+2
		5	195–222	28.0	+1	+5
		6	368–391	24.0	−8	−8
		7	403–422	20.0	−3	−8
am	Acm1_Human	1	23–51	29.0	−2	+4
		2	66–94	21.0	+4	+12
		3	107–125	19.0	+7	+4
		4	139–167	28.9	−3	+3
		5	184–212	29.0	−3	+3
		6	359–385	27.0	−8	−2
		7	Not predicted		(402–421)	
an	Acha_Human	1	232–260	29.0	−4	0
		2	269–293	25.0	+1	+7
		3	301–327	27.0	−1	+4
		4	428–456	29.0	−6	+4
b3	B3at_Human	1	400–426	27.0	−5	−2
		2	443–471	29.0	+9	0
		3	493–513	21.0	+2	0
		4	519–539	21.0	−4	−5
		5	569–593	25.0	−2	+5
		6	603–626	24.0	−1	+2
		7	657–685	29.0	−1	+5
		8	697–725	29.0	−9	−2
		9	768–796	29.0	+7	−10
		10	840–868	29.0	+7	−1
ba	Bacr_Halha	1	21–47	27.0	+2	+1
		2	57–83	27.6	+5	+6
		3	93–120	28.0	+4	+5
		4	120–143	24.2	+5	+3
		5	147–175	29.0	+4	+5
		6	187–207	21.0	+6	−1
		7	216–236	21.0	+3	−5
cb	Mcp1_Ecoli	1	3–31	29.0	−4	+1
		2	193–216	24.3	+2	+6
		3	361–382	22.0	f. p.	
cn	Cx43_Human	1	18–42	25.0	−2	+2
		2	70–96	27.0	−6	0
		3	151–176	26.4	−11	−5
		4	204–232	28.7	−3	+5

Table 4. Continued

Protein family	Reference sequence	Transmembrane segment				
		No.	Position	Length	Deviations	
co	Cox2–Human	1	21–46	27.0	–6	–2
		2	61–85	25.0	–2	+3
da	Dadr–Human	1	23–51	29.0	–4	+2
		2	60–88	29.0	–1	+1
		3	93–121	29.0	–5	+2
		4	136–164	28.4	–7	–16
		5	192–220	29.0	0	+5
		6	377–398	22.0	–1	–3
		7	Not predicted		(411–424)	
gb	Gaa1–Human	1	223–243	21.0	–1	–3
		2	251–271	21.0	+1	–1
		3	287–315	28.6	+3	+9
		4	393–414	22.0	–1	–1
gp	Glpb–Human	1	46–70	24.2	+2	+4
gt	Gtr1–Human	1	9–30	22.3	–4	–3
		2	67–87	21.0	0	0
		3	93–113	21.0	–3	–3
		4	117–145	28.9	–10	–2
		5	152–180	29.0	–4	+4
		6	186–214	29.0	0	+8
		7	268–292	25.0	–4	0
		8	310–330	21.0	+3	+3
		9	335–355	21.0	–3	–3
		10	364–392	29.0	–5	+3
		11	405–425	21.6	+3	+3
		12	432–460	28.6	+2	+10
ha	Hema–Ndvu	1	21–47	27.0	–6	–7
m1	Ha1a–Human	1	280–304	25.2	–5	–4
m2	Ha21–Human	1	194–222	29.0	–1	+5
m3	Hb2a–Human	1	195–223	29.0	–4	+4
mi	Mip–Bovin	1	10–35	27.7	–5	+1
		2	36–64	28.7	–5	+4
		3	77–105	29.0	–9	+1
		4	124–149	26.0	–7	0
		5	153–181	28.6	–4	+3
		6	202–225	24.0	0	+6
na	Nram–laudo	1	8–36	28.9	+1	+1
nr	Nram–Inbln	1	7–34	28.0	+6	–4

Table 4. Continued

Protein family	Reference sequence	Transmembrane segment				
		No.	Position	Length	Deviations	
op	Opsd_Human	1	35–63	29.0	–2	+2
		2	74–102	28.8	0	+3
		3	111–137	27.0	–3	+4
		4	149–176	28.3	–4	0
		5	202–230	29.0	–1	0
		6	249–275	27.0	–4	–1
		7	Not predicted		(285–309)	
ph	Psbb_Maize	1	17–41	25.0	–2	+1
		2	91–119	29.0	–4	+1
		3	137–165	29.0	–1	+2
		4	195–219	25.0	–2	0
		5	232–260	29.0	–4	0
		6	449–473	25.0	0	+5
ps	Psba_Pea	1	29–57	27.9	–7	+1
		2	105–130	26.8	–4	+1
		3	142–169	28.0	+1	+5
		4	169–194	26.0	f. p.	
		5	194–218	25.0	+2	0
		6	269–297	29.0	0	+8
rc	Rcem_Rhovi	1	47–75	27.2	–2	–2
		2	109–133	25.0	+2	–2
		3	140–163	24.0	0	–5
		4	194–222	29.0	+2	+2
		5	263–290	28.8	+1	0
t1	Atnb_Human	1	30–58	29.0	–5	–4
t2	Atpi_Pea	1	33–61	29.0	–7	–11
		2	93–121	29.0	–8	–4
		3	136–151	16.0	+2	–11
		4	189–209	21.0	+16	+12
		5	217–237	21.0	+1	–3
t3	Atha_Human	1	102–127	22.9	–1	+4
		2	136–161	23.8	–5	+1
		3	301–321	20.3	0	–1
		4	332–352	21.0	–2	–4
		5	810–838	29.0	–18	–8
		6	862–882	17.9	–1	–15
		7	956–984	28.7	–7	+1
		8	997–1023	22.2	–11	–1
t4	Atph_Pea	1	4–28	25.0	–3	–5
		2	48–71	24.0	–3	–6

Table 4. Continued

Protein family	Reference sequence	Transmembrane segment				
		No.	Position	Length	Deviations	
Pt5	Atn3–Human	1	82–102	21.0	–4	–5
		2	117–143	27.0	–2	+4
		3	283–308	26.0	+2	+4
		4	314–332	23.3	+4	+2
		5	774–802	29.0	–3	+2
		6	839–863	25.0	0	–1
		7	903–926	21.0	–3	0
		8	942–969	28.4	–1	+1

[a]For each family, the positions of the predicted transmembrane segments are related to a reference sequence, indicated by the Swissprot identifier. For comparison, deviations at the N- and C-termini are indicated relative to the calculated consensus from Swissprot or for the families ba, rc and t2 from Henderson *et al.* (1990), Yeates *et al.* (1987) and Hudson *et al.* (1987), respectively. For the three segments not predicted, the Swissprot positions are given within parentheses. The two false positive predictions are indicated by f.p. Lengths of predicted segments are the mean from the aligned sequences.

by recalculation of preferences with the respective exclusion of each family, followed by a prediction of the excluded familial member. In all the 28 families, the total mispredictions were still five. Thus, there is little bias due to the inclusion of all families and the information each generally provides is consistent. Given the large number (2319) of transmembrane proteins relative to the number of entries in each family (5–56), biases would not be expected.

10. Incorrect predictions

In the families of muscarinic acetylcholine receptors (family code, am), dopamine receptors (da) and opsins (op), the seventh helix is not predicted by the present method (*Table 4*). These families all belong to the superfamily of seven-helix G protein-coupled receptors (GPCRs) (Kerlavage, 1991). Other prediction algorithms also have problems with this family (*Table 5*) and in the same helical region which contains the retinyl-lysine and is adjacent to a potential acylation site (Savarese and Fraser, 1992)

There are also two cases of false positives. In the chemoreceptors family (cb), a third segment (positions 375–395) is predicted due to the high hydrophobicity of this region (*Table 4*). Analyses of enzyme activity in fusion proteins between alkaline phosphatase and different regions of these receptors show that the receptors have only two transmembrane segments (Manoil and Beckwith, 1986). In the photosystem Q(B) II D2 proteins (ps), similar reasons give rise to a false prediction of a transmembrane segment between helices 3 and 4 (*Table 4*). Sequence homologies with reaction centre proteins strongly suggest that only five transmembrane segments are present as observed in its tertiary structure where the region between helices 3 and 4 is constituted by a periplasmic helix with partial hydrophobic

Table 5. Comparisons of different prediction methods[a]

Protein family	Number of transmembrane segments	Mispredictions [average (min–max)]		
		TMAP	RA	KD
ad	7	0	1.3 (1–3)	1.0 (1–1)
am	7	1	2.5 (2–4)	2.1 (2–3)
an	4	0	0.1 (0–1)	0.0 (0–0)
b3	10	0	2.5 (2–4)	1.4 (1–4)
ba	7	0	0.8 (0–2)	1.3 (0–2)
cb	2	1	0.1 (0–1)	0.0 (0–0)
cn	4	0	0.0 (0–0)	0.5 (0–2)
co	2	0	0.1 (0–1)	0.0 (0–1)
da	7	1	1.0 (1–1)	1.0 (1–1)
gb	4	0	1.7 (1–3)	2.4 (1–5)
gp	1	0	0.0 (0–0)	0.0 (0–0)
gt	12	0	1.8 (0–4)	2.7 (1–5)
ha	1	0	0.5 (0–2)	0.1 (0–1)
m1	1	0	0.1 (0–1)	0.1 (0–1)
m2	1	0	0.0 (0–0)	0.1 (0–1)
m3	1	0	0.0 (0–0)	0.0 (0–0)
mi	6	0	0.7 (0–3)	1.5 (1–3)
na	1	0	0.1 (0–1)	0.0 (0–0)
nr	1	0	0.0 (0–0)	1.0 (1–1)
op	7	1	0.8 (0–1)	2.3 (1–4)
ph	6	0	0.0 (0–0)	2.7 (2–3)
ps	5	1	0.9 (0–2)	1.8 (0–4)
rc	5	0	0.6 (0–2)	1.3 (0–4)
t1	1	0	0.0 (0–0)	0.0 (0–0)
t2	5	0	2.0 (2–2)	1.0 (1–1)
t3	8	0	2.5 (1–4)	1.2 (0–3)
t4	2	0	0.0 (0–0	0.0 (0–0)
t5	8	0	1.9 (1–4)	2.3 (1–4)
Σ	126	5	22.0	27.8

[a]Results of predictions for transmembrane segments using the method in this work (TMAP) compared with the results from the techniques of Rao and Argos (1986) (RA) and Kyte and Doolittle (1982) (KD). Since the last two apply only to single sequences, the numbers in the table give averages with minimum and maximum values in parentheses. In ambiguous cases, predictions according to these two methods are interpreted favourably; for instance, if two nearly adjacent transmembrane helices were predicted as a long single span, a misprediction was not registered. Protein family abbreviations are as in *Table 3.*

character and lying atop the membrane with its axis parallel to the membrane plane (Deisenhofer and Michel, 1989).

Obviously, the quality of the alignments influences the predictions. It might therefore be helpful to consider the prediction graphs (*Figure 3*) for each alignment and check for consistency in $\langle P_m \rangle$ and $\langle P_e \rangle$ values; for instance, strong $\langle P_e \rangle$ peaks should sandwich a longer $\langle P_m \rangle$ peak. Clearly in ambiguous cases, the alignment should be checked for possible alterations.

11. Comparisons with other prediction methods

We compared the new technique with two widely used prediction methods: the method of Rao and Argos (1986) since it has been shown to be one of the most reliable of several methods (Degli Esposti *et al.*, 1990) and the technique of Kyte and Doolittle (1982). Since both these methods are designed to predict segments in single sequences, the comparative values were taken as the average number of mispredictions for each family as well as the minimum and maximum number of mispredictions over each family member (*Table 5*). The new algorithm is clearly more successful, with only five mispredictions over all families compared with 22 and 28 for the Rao/Argos and Kyte/Doolittle parameters, respectively. The increased information of the multiple sequence alignments and the hydrophobic and terminal propensity values are primarily responsible for the improvement (*Table 2*). The use of multiple sequence alignments to increase reliability has previously been described for secondary structure predictions (cf. Argos, 1988; Levin *et al.*, 1993; Persson *et al.*, 1991; *Rost et al.*, 1993).

12. Conclusions

A new method for the prediction of transmembrane amino acid sequence segments is presented. The method results in only five mispredictions in 126. Both the utilization of multiple sequence alignments and the use of two different sets of parameters for central and flanking regions contribute to the increased predictive reliability. With the addition of further sequences and more reliable 'standards-of-truth', the method can be further optimized. It would also be possible to extend the technique with the use of individual parametric sets for families of atypical membrane proteins.

Acknowledgements

This work was supported by the Alexander von Humboldt Foundation, the Swedish Medical Research Council (project no. 13F-10248), and the Swedish Society for Medical Research.

References

Argos P. (1988) A sequence motif in many polymerases. *Nucleic Acids Res.* **16**: 9909–9916.

Bairoch A. (1992) PROSITE: a dictionary of sites and patterns in proteins. *Nucleic Acids Res.* **20 (Suppl.)**: 2013–2018.

Bairoch A, Boekmann B. (1992) The SWISS-PROT protein sequence data bank. *Nucleic Acids Res.* **20 (Suppl.)**: 2019–2022.

Barnard EA. (1992) Receptor classes and the transmitter-gated ion channels. *Trends Biochem. Sci.* **17**: 368–374.

Capaldi RA. (1991) Structural features of the mitochondrial electron-transfer chain. *Curr. Opin. Struct. Biol.* **2**: 511–518.

Collins JF, Coulson AFW. (1990) Significance of protein sequence similarities. *Methods Enzymol.* **183**: 474–487.

Dayhoff MO, Barker WC, Hunt LY. (1983) Establishing homologies in protein sequences. *Methods Enzymol.* **91**: 524–545.

Degli Esposti M, Crimi M, Venturoli G. (1990) A critical evaluation of the hydropathy profile of membrane proteins. *Eur. J. Biochem.* **190**: 207–219.

Deisenhofer J, Michel H. (1989) The photosynthetic reaction centre from the purple bacterium *Rhodopseudomonas viridis*. *EMBO J.* **8**: 2149–2170.

Deisenhofer J, Epp O, Miki K, Huber R, Michel H. (1985) Structure of the protein subunits in the photosynthetic reaction centre of *Rhodopseudomonas viridis* at 3 Å resolution. *Nature*, **318**: 618–624.

Devereux J, Haberli P, Smithies O. (1984) A comprehensive set of sequence analysis programs for the VAX. *Nucleic Acids Res.* **12**: 387–395.

Etzold T, Argos P. (1993a) SRS — an indexing and retrieval tool for flat file data libraries. *CABIOS*, **9**: 49–57.

Etzold T, Argos P. (1993b) Transforming a set of biological flat file libraries to a fast access network. *CABIOS*, **9**: 59–64.

Griffith JK, Baker ME, Rouch DA, Page MGP, Skurray RA, Paulsen IT, Chater KF, Baldwin SA, Henderson PJF. (1992) Membrane transport proteins: implications of sequence comparisons. *Curr. Opin. Struct. Biol.* **4**: 684–695.

von Heijne G. (1986) The distribution of positively charged residues in bacterial inner membrane proteins correlates with the trans-membrane topology. *EMBO J.* **5**: 3021–3027.

von Heijne G. (1992) Membrane protein structure prediction. Hydrophobicity analysis and the positive-inside rule. *J. Mol. Biol.* **225**: 487–494.

Henderson R, Baldwin JM, Ceska TA, Zemlin F, Beckmann E, Downing KH. (1990) Model for the structure of bacteriorhodopsin based on high-resolution electron cryo-microscopy. *J. Mol. Biol.* **213**: 899–929.

Henikoff S, Henikoff JG. (1992) Amino acid substitution matrices from protein blocks. *Proc. Natl Acad. Sci. USA*, **89**: 10915–10919.

Hudson GS, Mason JG, Holton TA, Koller B, Cox GB, Whitfeld PR, Bottomley W. (1987) A gene cluster in the spinach and pea chloroplast genomes encoding one CF_1 and three CF_0 subunits of the H^+-ATP synthase complex and the ribosomal protein S2. *J. Mol. Biol.* **196**: 283–298.

Iwata S, Ostermeir C, Ludwig B, Michel H. (1995) Structure at 2.8 Å resolution of cytochrome *C* oxidase from *Paracoccus denitrificans*. *Nature*, **376**: 660–669.

Jennings ML. (1989) Topography of membrane proteins. *Annu. Rev. Biochem.* **58**: 999–1027.

Kerlavage AR. (1991) G-protein-coupled receptor family. *Curr. Opin. Struct. Biol.* **1**: 394–401.

Krauss N, Hinrichs W, Witt I, Fromme P, Pritzkow W, Dauter Z, Betzel C, Wilson KS, Witt HT, Saenger W. (1993) Three-dimensional structure of system I of photosynthesis at 6 Å resolution. *Nature*, **361**: 326–331.

Kühlbrandt W, Wang DN, Fujiyoshi Y. (1994) Atomic model of plant light-harvesting complex by electron crystallography. *Nature*, **367**: 614–621.

Kyte J, Doolittle RF. (1982) A simple method for displaying the hydropathic character of a protein. *J. Mol. Biol.* **157**: 105–132.

Lemmon MA, Engelman DM. (1992) Helix–helix interactions inside lipid bilayers. *Curr. Opin. Struct. Biol.* **2**: 511–518.

Levin JM, Pascarella S, Argos P, Garnier J. (1993) Can multiple alignments improve secondary structure predictions? *Protein Eng.* **6**: 849–854.

Li J. (1992) Bacterial toxins. *Curr. Opin. Struct. Biol.* **2**: 545–556.

McDermott G, Prince SM, Freer AA, Hawthornthwaite-Lawless AM, Papiz MZ,

Cogdell RG, Isaacs NW. (1995) Crystal structure of an integral membrane light-harvesting complex from photosynthetic bacteria. *Nature*, **374**: 517–521.

Manoil C, Beckwith J. (1986) A genetic approach to analyzing membrane protein topology. *Science*, **233**: 1403–1408.

Marger MD, Saier Jr MH. (1993) A major superfamily of transmembrane facilitators that catalyse uniport, symport and antiport. *Trends Biochem. Sci.* **18**: 13–20.

Miller C. (1991) 1990: Annus mirabilis of potassium channels. *Science*, **252**: 1092–1096.

Needleman SB, Wunsch CD. (1970) A general method applicable to the search for similarities in the amino acid sequence of two proteins. *J. Mol. Biol.* **48**: 443–453.

Persson B, Argos P. (1994) Prediction of transmembrane segments in proteins utilising multiple sequence alignments. *J. Mol. Biol.* **237**: 182–192.

Persson B, Krook M, Jörnvall H. (1991) Characteristics of short-chain alcohol dehydrogenases and related enzymes. *Eur. J. Biochem.* **200**: 537–543.

Picot D, Loll PJ, Garavito RM. (1994) The X-ray crystal structure of the membrane protein prostaglandin H_2 synthase-1. *Nature*, **367**: 243–249.

Popot J-L, de Vitry C. (1990) On the microassembly of integral membrane proteins. *Annu. Rev. Biophys. Biophys. Chem.* **19**: 369–403.

Rao JKM, Argos P. (1986) A conformational preference parameter to predict helices in integral membrane proteins. *Biochim. Biophys. Acta*, **869**: 197–214.

Rost B, Schneider R, Sander C. (1993) Progress in protein structure prediction. *Trends Biochem. Sci.* **18**: 120–123.

Savarese TM, Fraser CM. (1992) In vitro mutagenesis and the search for structure–function relationships among G protein-coupled receptors. *Biochem. J.* **283**: 1–19.

Schloss P, Mayser W, Betz H. (1992) Neurotransmitter transporters. A novel family of integral plasma membrane proteins. *FEBS Lett.* **307**: 76–80.

Sipos L, von Heijne G. (1993) Predicting the topology of eukaryotic membrane proteins. *Eur. J. Biochem.* **213**: 1333–1340.

Stephenson FA. (1991) Ion channels. *Curr. Opin. Struct. Biol.* **1**: 569–574.

Thompson JD, Higgins DG, Gibson TJ. (1994) CLUSTAL W: improving the sensitivity of progressive multiple sequence alignment through sequence weighting, position specific gap penalties and weight matrix choice. *Nucleic Acids Res.* **22**: 4673–4680.

Traxler B, Boyd D, Beckwith J. (1993) The topological analysis of integral cytoplasmic membrane proteins. *J. Membrane Biol.* **132**: 1–11.

Unwin N. (1993) Nicotinic acetylcholine receptor at 9 Å resolution. *J. Mol. Biol.* **229**: 1101–1124.

Vingron M, Argos P. (1989) A fast and sensitive multiple sequence alignment algorithm. *CABIOS*, **5**: 115–121.

Weiss MS, Schulz, GE. (1992) Structure of porin refined at 1.8 Å resolution. *J. Mol. Biol.* **227**: 493–509.

Yeates TO, Komiya H, Rees DC, Allen JP, Feher G. (1987) Structure of the reaction center from *Rhodobacter sphaeroides* R-26: Membrane–protein interactions. *Proc. Natl Acad. Sci. USA*, **84**: 6438–6442.

Chapter 2

Sequence–function correlation in G protein-coupled receptors

W. Kuipers, L. Oliveira, A.C.M. Paiva, F. Rippmann, C. Sander, G. Vriend, A.P. IJzerman

1. Introduction

One of the central paradigms in protein research is that the sequence determines the structure, and the structure the function. Many G protein-coupled receptor (GPCR) sequences have been determined, and much functional data are available. Structural data, however, are very limited (Schertler *et al.* 1993; Unger and Schertler, 1995), and modelling studies have not yet yielded adequately accurate models to allow for the inference of function. It would therefore be useful if the pathway sequence to function could be studied without the need for structure determination or modelling studies. Multiple sequence alignments provide a potentially powerful tool for such analyses (Casari *et al.*, 1994).

Ever since Eck and Dayhoff (1969) suggested how evolutionary changes in proteins could be described by a residue exchange matrix, the question of which matrix is the best has remained under debate. Eck and Dayhoff described a matrix that indicates the likelihood of mutations occurring during evolution. This matrix can be modified for use as a scoring matrix in sequence alignment procedures. In this case, the matrix is used to assess the likelihood that two residues occur at equivalent positions in a sequence alignment. However, the same mutations are not equally likely to occur at different positions in a sequence, and using a single scoring matrix for all positions in the sequences to be aligned is often inadequate. Overington *et al.* (1992) extended Eck and Dayhoff's idea by using multiple matrices; one for residues in a helix, another for residues in a beta strand, etc. Scharf (1989) and Bowie *et al.* (1991) went one step further and created one exchange matrix for each position in the sequence. These so-called structure based profiles can of course only be constructed if a three-dimensional structure for at least one of the family members is available. Sander and Schneider (1991) described structure-based multiple sequence alignments, which are sequence profiles for protein structures that are produced from multiple sequence alignments.

GPCRs reside partly in an aqueous environment and partly embedded in a lipid membrane. Thus, very different physico-chemical constraints apply at different positions in the sequence. Consequently, evolutionary pressure is different for different positions, and a single Dayhoff-type matrix is inadequate for GPCR sequence alignment. We therefore used an iterative sequence alignment similar to the one described by Sander and Schneider (1991). Rather than using a scoring matrix, we made the profile relate directly to the frequencies of the residues at each position in the multiple sequence alignment.

The main aim of most GPCR modelling studies in medicinal research is the design of (specific) agonists or antagonists. For these studies, knowledge of the residues involved in ligand binding is very important. Much indirect evidence points to the idea that the same drug often binds in a similar manner in structurally related receptors (Savarese and Fraser, 1992). In these cases, a correlation between the presence or absence of certain key receptor residues and the affinity for a specific drug is to be expected.

Biogenic amine receptors are the best studied subclass of GPCRs. The experimental receptor binding data that emerged from these studies form the basis for our analyses in which we perform a correlated mutation analysis between ligand affinities and receptor sequences. Mutagenesis data enable examination of the results of these analyses.

For many GPCRs, residues have been experimentally identified as possibly being involved in agonist binding. In constrast to all other GPCRs, opsins are not activated by an agonist, but by a photon that triggers a cis–trans isomerization in the tightly bound retinal. This retinal is located between the transmembrane helices, in a similar position to that thought to be occupied by small ligands in many other GPCRs. It seems probable that the spectral characteristics of rhodopsin are in part determined by residues in the vicinity of the retinal.

Here, we present a simple method to analyse sequence–function relationships in GPCRs. The method searches for correlated mutations in multiple sequence alignments. This method was used to search for residues that determine the optimal wavelength for retinal activation in opsins, and to analyse which residues are important for agonist or antagonist binding in serotonin receptors and adrenoceptors.

2. Data and methods

Sequences

In this discussion one numbering scheme has been adopted for all sequences analysed. *Figure 1* shows a multiple sequence alignment including the consenus sequence for each subclass of GPCRs in this study. A unified numbering scheme is indicated, similar to the one used by Hibert *et al.* (1991).

To enhance the signal-to-noise ratio in our analyses we used as many sequences as possible including species variants. In a few cases, the sequence differences between very unrelated species (e.g. mammals and insects) were too large, and the study was restricted to mammals.

Sequences were extracted from the Swissprot database (Bairoch, A. Department Biochimie Medicinale, Centre Medicinal Universitaire, 1211, Geneva 4, Switzerland, and SWISS-PROT Protein sequence database, EBML Data Library, D-69117 Heidelberg, Germany), the PIR database (PIR, NBRF, Georgetown University Medical Center, 3900 Reservoir Road NW, Washington, DC), the translated GenBank (NCI-FCDRDC, Frederick Biomedical Supercomputing Center, PO Box B, Frederick, MD 21702-1201.) and from the GCRDb database (Kolakowsky, 1994). The program WHAT IF (Vriend, 1990, available from G.V.) was used for database searches, multiple sequence alignments and correlated mutation analyses.

Iterative unbiased profile alignment of sequences

We used a profile-based alignment method similar to that described by Sander and Schneider (1991). All biases that could be introduced by the use of a scoring matrix were removed by making profiles relate directly to the frequency of occurrence of residue types at each position in the protein. Sequences were aligned against this profile, rather than against all other sequences,

This now creates a problem: in order to do the alignment, we need a profile, and in order to get a profile, we need an alignment. To solve this problem a multiple sequence alignment was constructed on a subset of the sequences that show high pairwise homologies. From this initial alignment a profile was created. This profile was then used to align all the sequences of interest. The aligned sequences were sorted according to their similarity to the profile, and a new profile was made from the highest scoring sequences. This process was repeated until all sequences were incorporated. In every iteration more sequences were incorporated than in the previous one. Sequences that show too little similarity with the consensus sequence of the profile were not used because there is no guarantee that they belong in the same structural family. After incorporating all sequences of interest the alignment procedure was iterated a few more steps. In practice, this iterative alignment method normally converged to a satisfactory solution for multiple sequence alignments in less than five cycles, provided enough sequences were available (typically 20–30 sequences are enough).

```
                        Extracellular                   Cytosolic
          I            111. . . . . . . .120. . . . . . . . . .130. . . .135

Bacteriorhodopsin       P E W I W L A L G T A L M G L G T L Y F L V K G M
                          *       *     *       *     *       *       *
Alpha adrenoceptor      A T A G L A L L L G L L I L F T V F G N V L V I L
Beta adrenoceptor       W V A G M G L L M A L I V L A I V F G N V L V I T
Dopamine                H Y N Y Y A C L L T L L I L V I L F G N V L V C A
Muscarinic              E V V F I A T V T G S L S L V T V I G N I L V M V
Serotonin               E K V W V A L L L G L I I L L T I A G N I L V I A
(Rhod)opsin             V Y S L L A A Y M F F L I I I G F P G N G L V L Y
Peptide                 L A V L I P A L Y S V V F V V G L L G N S L V I W
Protein                 G Y K I L R I V Y W F V F L L S L L G N I L V L L
          II
                             240. . . . . . . . .230. . . . . . . . .220. . .216

Bacteriorhodopsin             L L M S L Y M T F A I A P V L T T I A Y F K K A D
                                  *     *       *     * *     *       *
Alpha adrenoceptor            L E N A A S F P I V L T A V L L D A S A L S V I F
Beta adrenoceptor             L I H T A G F P V V L L G M V L D A C A L S T I F
Dopamine                      V E L Y V V W P M V L T A V L L D A V A L S V I L
Muscarinic                    I I Y T T Y L N M S F V G I I L D A C A L S F L F
Serotonin                     L I Y L L S L P M V L V A V L L D A V A L S M I L
(Rhod)opsin                   L S N Y M T T T F G G L V M F L D A V A L N V L I
Peptide                       L A Y Y V A W F P L T L A F L L D A L A L N L I F
Protein                       I L D V S A I L L L Y L G F L L D A F A L N C M L


          III         314. . . . .320. . . . . . . . .330. . . . . . . . . .340.

Bacteriorhodopsin     E Q N P I Y W A R Y A D W L F T T P L L L L D L A L L .
                      * *     * *     * *   * *     * *   * *     * *   *
Alpha adrenoceptor    W C E I W L A L D V L C C T A S I V H L C A I S L D R Y
Beta adrenoceptor     W C E L W T S V D V L C V T A S I E T L C V I A V D R Y
Dopamine              H C D I F V T L D V M M C T A S I L N L C A I S I D R Y
Muscarinic            V C D L W L A L D Y V A S N A S V M N L L I I S F D R Y
Serotonin             L C D V W I S L D V L C C T A S I M H L C A I A L D R Y
(Rhod)opsin           G C N I E G F F G S L G G I V A L W S L V V I A I E R Y
Peptide               L C K L V N F I D G T N M F T S I F S L T A L S V D R Y
Protein               L C N V A G F L T V F S S Y A S V Y T L T V I T L D R Y
          IV           432.430. . . . . . . . .420. . . . . . . .411

Bacteriorhodopsin       L A G V L G T G I M I G D A G V L A L I T G
                          * *     * *     *     * *     *
Alpha adrenoceptor      I L P P F S I V A S I V W V T L I I A K A R
Beta adrenoceptor       Q I P L F S V L A S V I W V M L V A A R A R
Dopamine                F L L P C S I A F S L V W V I Q I M V T V R
Muscarinic              L I A P A W L V F S L V W A A A I M L G A M
Serotonin               F L P P L S I G I S I L W V I L I M L A A R
(Rhod)opsin             F L P P A A W A L A M I W T F A I G M I A H
Peptide                 I A Q P L S L L L A L V W I V V I V V K A V
Protein                 V L P P L A L L I S F V W I G L M I F K A R


          V                           508.510. . . . . . . . .520. . . . . . .528

Bacteriorhodopsin                     W W A I S T A A M L Y I L Y V L F F G F T
                                        *     * *     *     * *     * *   * *
Alpha adrenoceptor                    Y A L A S S I G S F Y A P L L I M L L V Y
Beta adrenoceptor                     Y A I A S S I V S F Y V P L L V M V F V Y
Dopamine                              F V I Y S S I V S F Y V P F I V T L L V Y
Muscarinic                            I T F G T A I A A F Y L P V T I M T V L Y
Serotonin                             Y T I Y S T F G A F Y I P L T L M L I L Y
(Rhod)opsin                           Y V I Y M F I F H F I I P L I I I F F C Y
Peptide                               Y G I L L F L L G F L L P L L I I A V C Y
Protein                               L Y I L F F L V L F I V P L L I I L I C Y

          VI                    627. . . . . .620. . . . . . . . . .610. . . . . . .602

Bacteriorhodopsin               I L W V V P Y A S W L V V T V N R L V K F T S A V E
                                    *     * *     *     * *     * *   * *     * *
Alpha adrenoceptor              L P N A F F F P L W C L V F V G M V I G L T K T A K
Beta adrenoceptor               V V N A L F F P L W C L T F T G M I I G L T K L A K
Dopamine                        L I H T I F F P L W C I I F V G L V I A L M Q T A K
Muscarinic                      V L V M I N Y P T W T I I F A L L I A S I T R A A K
Serotonin                       V L N T I F F P L W C V V F A G F I I G L T K A A K
(Rhod)opsin                     Y F A V V A Y P T W C I L F A I V M V I V M R T V E
Peptide                         L L N F I H F P L W C L A F V A V V V L V M K T V R
Protein                         S I L L F S Y P A W C L F F T F I L A A M M K A I R
          VII             .710. . . . . . . . . .720. . . . . . . . . .730.732

Bacteriorhodopsin         N I E T L L F M V L D V S A K V G F G L I L L R
                              * *     * *     *     * *     * *     *
Alpha adrenoceptor        L V P E G L F K F F F W L G Y C N S S L N P V I
Beta adrenoceptor         L V P K E V F I L L N W L G Y A N S A F N P L I
Dopamine                  N I P P V L Y S A F T W L G Y V N S A L N P I I
Muscarinic                C I P E T V W S I G Y W L C Y V N S T I N P A C
Serotonin                 C I P L A L L N F F L W L G Y L N S L I N P L I
(Rhod)opsin               A F T P L A M T I P A F F A K S S A I Y N P I I
Peptide                   K F V D T V Y L I T I A L A Y A N S C L N P I L
Protein                   I T V T N A K I L L V L F Y P L N S C L N P I L      *1)
```

Analysis of correlated mutations

Correlated mutational behaviour can be defined as the tendency of residues to stay conserved or to mutate in tandem between (sets of) sequences. Several techniques have been described to analyse correlated mutations. Goebel *et al.* (1994), for example, found that the correlation coefficient is related to the chance that the residue pair is in contact in the structure. In this method, completely conserved residues give rise to a high correlation, indicating their importance for structural integrity. This procedure works well if structural analysis or structure prediction is the primary aim. To determine functional correlations, however, a slightly different approach is required (Casari *et al.*, 1994; Oliveira *et al.*, 1993). Oliveira *et al.* (1993) used a similar strategy to Goebel *et al.* (1993) to determine the correlation between residue positions in GPCRs. It requires a certain degree of variability at positions that are potentially involved in behavioural characteristics. *Figure 2* shows an example.

Using this method many functional residues in GPCRs were detected (Oliveira *et al.*, 1993), and a hypothesis for the activation pathway could be formulated (Oliviera *et al.*, 1994). However, rather than determining whether residue positions show pairwise correlated mutational behaviours, residues can also be correlated with protein-related parameters. Parameters that can be used are, for example, the binding constant for a certain ligand or the subclass of receptors to which the sequence belongs to. These parameters can be encoded in a single character called a pseudo residue. This pseudo residue can be used in the correlation analysis as if it is a normal residue.

We have explored several simple scoring schemes to determine correlations between pseudo residues and real residues. *Figure 3* shows the scheme that appeared to be most useful when analysing functional characteristics, for instance endogenous ligand binding. This scheme is very similar to the one described in *Figure 2*. The main difference is that residue positions are not compared pairwise, but only with the pseudo residue. A residue (see *Figure 3*) is called uncorrelated if it is different from the majority of residues that correspond with the same pseudo residue. If it is the same as the majority of residues corresponding to another pseudo residue, it is called anti-correlated.

A different approach was chosen for the analysis of exogenous ligand binding. In contrast to binding of endogenous ligands, the binding of exogenous ligands is not a natural property of the receptor. The objective using this approach was to find residues that are conserved in the receptors that bind the exogenous ligand, but are

---

Figure 1. Alignment of the consensus sequences of several GPCR subclasses on the helices of bacteriorhodopsin (Henderson *et al.*, 1990). This alignment is based on that suggested by Oliveira *et al.* (1993).

```
Sequence    5    10
            |    |                          Sequence position
1       AAAASSSSTTTT                        Positions 2 and 3 are compared with this one
2       RRRRPPPPHHHH   66 1.00              All residues correlate perfectly with position 1
3       TTTTGGGGEEED   63 0.95              One residue is not correlated with position 1
```

Figure 2. Hypothetical example of correlated mutations. In this multiple sequence alignment two residue positions are compared with the first position. All residue pairs are compared, leading to 12(12 – 1)/2 = 66 comparisons. If the residues are either conserved or mutated in both sequences, a score of 1 is given. The correlation between a residue pair is defined as the score divided by the maximal score (66). The sequences move down the page, so three sequence positions are given in twelve sequences.

```
        5    10
        |    |
-1  111122223333              Pseudo residue
 1  RRRRPPPPHHHH   12 1.00    All residues correlate with the pseudo residue
 2  TTTTGGGGEEED   11 0.92    The D does not correlate
 3  LLLLAAAAYLYY   10 0.84    The L is anti-correlated
```

Figure 3. Hypothetical example of scoring correlated mutations. In this multiple sequence alignment, three residue positions 1, 2 and 3 are compared with a pseudo residue (–1). If two residues are the same and the corresponding pseudo residues are the same, a score of 1 is given. If a residue is not correlated with the pseudo residue the score is 0. Anti-correlated residues score -1. The correlation between a residue pair is defined as the score divided by the maximal score (12). The sequences run down the page, so three sequence positions are given in twelve sequences.

absent in the receptors to which the exogenous ligand does not bind. In receptors with low affinity, any residue is allowed, provided it is different from that in receptors with high affinity. A very simple scoring scheme can be used to find such residues. Receptors that bind the ligand well are given pseudo residue '+'; and receptors with low affinity '–'. For receptors with pseudo residue '+', a residue has to be identical to the majority of the '+' coded residues in the same position, in which case a score of 1 is given. If a residue in the '–' coded receptors belongs to this same majority the score is –1. The final score is divided by the maximal achievable score.

Within a given data set the selectivity of a compound may depend on the absence or presence of more than just one residue. For example, binding can be abolished for steric reasons if an alanine is mutated into a leucine, or for energetic reasons if a hydrogen bond is lost because of a threonine to valine mutation. Thus a scoring scheme was designed that reflects a dependence on a combination of residues. In this scheme individual residues that are involved in binding may also be present in '–' coded receptors, but for every pair of positions at least one should not be the

Sequence number	Seq. pos. 1	2	3	4	5	6	7	8	9	10	Compound 1	2	3
1	**F**	I	A	V	**H**	F	A	A	A	G	+	+	–
2	**F**	I	V	V	**H**	G	**S**	A	A	G	+	+	+
3	**F**	V	I	D	**H**	S	**S**	G	A	G	+	+	+
4	G	I	L	R	**H**	T	**S**	G	A	G	–	–	–
5	**F**	I	Y	I	**H**	V	T	S	A	F	+	+	–
6	**F**	I	G	L	W	V	A	S	A	F	+	–	–
7	**F**	A	T	C	W	W	A	T	A	F	+	–	–
8	G	S	I	C	W	W	**S**	V	A	L	–	–	–
9	**F**	I	A	S	W	I	**S**	I	A	L	+	–	–
10	**F**	I	C	T	W	L	**S**	L	L	L	+	–	–

Figure 4. A hypothetical set of 10 sequences of 10 residues each. The affinities of three hypothetical compounds for each receptor are coded with pseudo-residue '+' for high affinity, and '–' for low affinity. Binding of compound 1 is correlated only with the presence of a phenylalanine (**F**) at position 1. Our simplest scorings scheme detects such cases. Compound 2 correlates with the presence of a phenylalanine (**F**) at position 1 *and* a histidine at position 5. The scoring scheme that is based on a comparison of pairs of residue positions with the pseudo residue will detect such cases. Compound 3 depends on the combined presence of phenylalanine (**F**) 1, histidine (**H**) 5 and serine (**S**) 7. At present we cannot detect such cases.

same as the majority of the '+' coded residues. A hypothetical example explaining the different scoring schemes is given in *Figure 4*.

3. Results and discussion

Opsin residues determining the wavelength of photons absorbed by retinal

The optimal wavelength of an incoming photon is determined by the precise chemical environment of the retinal, and it can therefore be assumed that there are some residues that are important for determining the optimal wavelength of the absorbed photon. Recently, it has been shown experimentally that His-440 and Lys-443 (note special numbering arrangement extracellular in the loop between helices IV and V) play a role in the light absorption of green and red visual pigments (Wang *et al.*, 1993). These two charged residues could form a chloride binding site. From the presence of a disulphide bond between the cysteine in loop IV-V and Cys-315 at the external side of helix III (Karnik and Khorana, 1990) it can be inferred that His-440 and Lys-443 are positioned in the vicinity of Glu-318, the residue assumed to be a counterion for the retinylidene Schiff's base (Zhukovsky and Oprian, 1989; Sakmar *et al.*, 1989, 1991). Under physiological conditions, chloride ions that are present in the extracellular medium saturate this binding site, leading to a red shift in the retinal absorption maximum. In the absence of those two residues (especially His-440), red or green pigments behave like other shorter wavelength absorbing rhodopsins (Wang *et al.*, 1993).

To determine which residues influence the optimal wavelength in vertebrate opsins, we looked for all residue positions that display perfect correlated mutational behaviour (*Table 1*). Nine residue positions form a network of perfect pairwise correlations. At these positions all blue/violet opsins and the green pigments of goldfish and chicken have the same residue type whereas the other opsins (red/green) systematically have another conserved residue at this position. One of the two residues from the chloride site (Lys-443) is part of this network. His-440 is not perfectly correlated because sheep rhodopsins have a glutamine instead of a glutamic acid at this position. The sequences used in this study can be obtained from the TM7 file server (WWW address is http://swift.EMBL-Heidelberg.DE/).

Table 1. Wavelength determining residues in opsins[a]

	103	126	337	443	444	452	517	531	738
Blue/violet	Q	G	A	Q	C	T	F	L	K
Red/green	N	S	S	K	T	S	C	V	R

[a]All residues that form a perfect network of correlated mutations are shown.

Residues important for ligand specificity in biogenic amine receptors

Many receptors, such as serotonin (5-HT), adrenergic, muscarinic, dopamine, and histamine receptors, interact with ligands that contain a positively charged nitrogen atom. Several of the endogenous agonists (neurotransmitters) for these receptors are shown in *Figure 5*. The structure of acetylcholine, which activates muscarinic receptors, is rather different from other aminergic neurotransmitters. The aromatic ring system, which is present in most endogenous amine agonists, is replaced by a polar non-aromatic acetyl group in acetylcholine. Furthermore, it contains a quaternary ammonium group with three methyl substituents instead of a primary or secondary amine group as in other aminergic neurotransmitters. This raises the question whether any residues exist that correlate with these structural differences. We compared 32 muscarinic receptor sequences with 144 other biogenic amine receptor sequences. The seven residues that discriminate best between these two classes are shown in *Table 2*. They are all located in or close to the putative ligand binding pocket for biogenic amine receptors. Residue positions 327 and 330 are in the middle of helix III close to the conserved aspartate residue (Asp-322) that interacts with the ligand's ammonium group. Residue position 231 is near Asp-322 in most GPCR models. The conserved mutations in positions 621 and 622, in the middle of helix VI, are of particular interest. At position 621, a highly conserved Phe in serotonin, dopamine and adrenergic receptors, is replaced by an equally conserved Tyr in muscarinic and histamine receptors. The more polar character of Tyr, and its capability of forming hydrogen bonds, is in good agreement with the more polar character of their endogenous agonists.

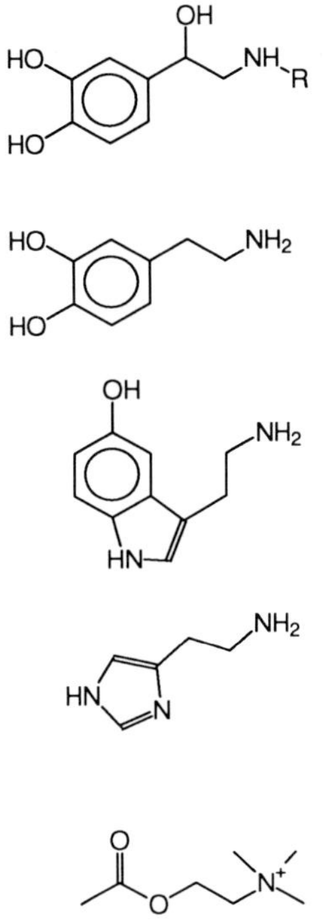

Noradrenaline R=H
Adrenaline R=CH_3

Dopamine

Serotonin

Histamine

Acetylcholine

Figure 5. Natural agonists for amine receptors.

Table 2. Residues that discriminate between muscarinic and other biogenic amine receptors[a]

	231	233	327	330	621	622	722
Muscarinic	S	N	N	V	Y	N	C
Other amine	V	P	T	I	F	F	G
	(5I)	(1S)		(1H)	(1A)		

[a]A list of the 176 sequences used can be obtained from the TM7 file server. Thirty-two muscarinic receptors were compared with 144 other amine receptors. Four of these residue positions are not 100% conserved in the non-muscarinic aminergic receptors. The alternative residues and their frequency of occurrence are indicated between brackets.

The importance of residue 621 for muscarinic agonist affinity was confirmed by the mutation of Tyr-621-Phe in M3 muscarinic receptors, which decreases agonist binding (Wess *et al.*, 1992). The highly conserved Phe-622 in all other amine receptors is an Asn in all muscarinic receptors. This mutation agrees with the structural differences between the endogenous agonists of these receptors. Asn-622 is probably capable of forming hydrogen bonds with the polar acetylcholine, whereas Phe-622 has an aromatic–aromatic interaction with the other amine neurotransmitters. The importance of Phe-622 for agonist affinity was confirmed by mutation studies in the β-adrenoceptor and the 5-HT_{2A} receptor (Strader *et al.*, 1989;

Choudhary *et al.*, 1993). Residue 722 is adjacent to positions claimed to bind acetylcholine in muscarinic receptors (Wess *et al.*, 1992).

The aryloxypropanolamine binding site

Pindolol and propranolol are β-adrenoceptor antagonists, belonging to the class of the aryloxypropanolamines (*Figure 6*). These antagonists are not selective for this class as they display considerable affinity for rodent 5-HT_{1B} and all 5-HT_{1A} receptors (Boess and Martin, 1993). To identify the residue(s) important for binding compounds of this class, we analysed affinities for pindolol and propranolol. Because these compounds were not always tested on the same receptors, two different sets of receptor-affinity data were used (*Table 3*). All receptors with high affinity (all β_1 and β_2, 5-HT_{1A} and rodent 5-HT_{1B}) were given pseudo residue '+', and those having low affinity were given pseudo residue '–'. *Figure 7* shows that Asn-719 displays a 100% correlation with affinity for both compounds despite the fact that two different sets of receptor-affinities were used.

The importance of Asn-719 for binding pindolol and propranolol has been confirmed in a number of cases. The mutation Asn-719-Val in the 5-HT_{1A} receptor, decreases the receptor's affinity for pindolol, but has no effect on the affinity for the natural agonist (serotonin) (Guan *et al.*, 1992). The Phe-719-Asn mutation in the α_2-adrenoceptor increases affinity for these antagonists considerably (Suryanararayana *et al.*, 1991). The Thr-719-Asn mutation in human 5-HT_{1B} and 5-$HT_{1D\beta}$ receptors increases their affinities for pindolol and propranolol. In these last two studies, it was shown that the Thr versus Asn difference in position 719 was completely responsible for the receptor-binding differences between rodent 5-HT_{1B} receptors and other 5-HT_{1B} and 5-HT_{1D} receptors (Parker *et al.*, 1993; Oksenberg *et al.*, 1992).

Binding site of 5-carboxamidotryptamine in serotonin receptors

Serotonin displays high affinity for almost all serotonin receptors (for review see Boess and Martin, 1993). Replacement of the 5-hydroxy group of serotonin with a carboxamido group, yielding 5-carboxamidotryptamine (5-CT, see *Figure 6*), increases the affinity for 5-$HT_{1A,B,D}$, 5-HT_5 and 5-HT_7 receptors by a factor of 5–10 (Boess and Martin, 1993). In contrast, the affinity for 5-$HT_{1E,F}$, 5-HT_2 and 5-HT_6 receptors decreases by a factor of 10–100. Like serotonin, 5-CT was shown to act as an agonist on 5-HT_{1A} receptors, suggesting a similar binding mode for both compounds.

We performed a sequence analysis, using pseudo residue '+' for 5-$HT_{1A,B,D}$, 5-HT_5 and 5-HT_7 receptors, and '–' for 5-$HT_{1E,F}$, 5-HT_2 and 5-HT_6 receptors (*Table 4*).

Table 3. Affinities for propranolol and pindolol, represented by pseudo residues, and receptor sequence which were used for correlation the analysis

Accession number	Pseudo residues[a] Pindolol	Propranolol	Receptor type	Species
P19327	+	+	Serotonin 1A	Rat
P08908	+	ND	Serotonin 1A	Human
P28564	+	+	Serotonin 1B	Rat
P28334	+	ND	Serotonin 1B	Mouse
P07700	+	+	Beta-1 adrenoceptor	Turkey
P18090	+	+	Beta-1 adrenoceptor	Rat
P10608	+	+	Beta-2 adrenoceptor	Rat
P28222	–	–	Serotonin 1B	Human
P28565	–	–	Serotonin 1D	Rat
P28221	–	–	Serotonin 1D	Human
P28566	–	ND	Serotonin 1E	Human
P30939	–	ND	Serotonin 1F	Human
P30940	–	ND	Serotonin 1F	Rat
Q02284	–	ND	Serotonin 1F	Mouse
P14842	–	–	Serotonin 2A	Rat
P08909	–	ND	Serotonin 2C	Rat
P30966	–	–	Serotonin 5A	Mouse
P35364	–	–	Serotonin 5A	Rat
P31387	–	–	Serotonin 5B	Mouse
P35365	–	–	Serotonin 5B	Rat
P31388	–	ND	Serotonin 6	Mouse
P32304	ND	–	Serotonin 7	Mouse
P32305	ND	–	Serotonin 7	Rat
P34969	ND	–	Serotonin 7	Human
P20288	ND	–	Dopamine 2	Bovine
P13953	–	–	Dopamine 2	Rat
P23944	ND	–	Alpha-1A adrenoceptor	Rat
P15823	ND	–	Alpha-1B adrenoceptor	Rat
P18130	ND	–	Alpha-1C adrenoceptor	Bovine
P08913	ND	–	Alpha-2A adrenoceptor	Human
P22909	ND	–	Alpha-2A adrenoceptor	Rat
P18089	ND	–	Alpha-2B adrenoceptor	Human
P19328	ND	–	Alpha-2B adrenoceptor	Rat
P18825	ND	–	Alpha-2C-1 adrenoceptor	Human
P22086	ND	–	Alpha-2C adrenoceptor	Rat
P35369	ND	–	Alpha-2C-2 adrenoceptor	Human
P08482	–	–	Acetylcholine 1	Rat
P10980	–	–	Acetylcholine 2	Rat
P08483	–	–	Acetylcholine 3	Rat
P08485	–	–	Acetylcholine 4	Rat
P08911	–	–	Acetylcholine 5	Rat
P31389	ND	–	Histamine 1	Guinea pig
P35367	ND	–	Histamine 1	Human
P31390	ND	–	Histamine 1	Rat

[a]Receptor affinity values were taken from Boess and Martin (1993) and Seeman (1993). All receptors with high affinity have $pK_i \geq 7.0$ and were coded '+'. The receptors with low affinity all have $pK_i < 6.0$ and were coded '–'.
ND indicates that no data are available. Twenty-seven sequences were used in the pindolol analysis; 36 sequences were used in the propranolol analysis.

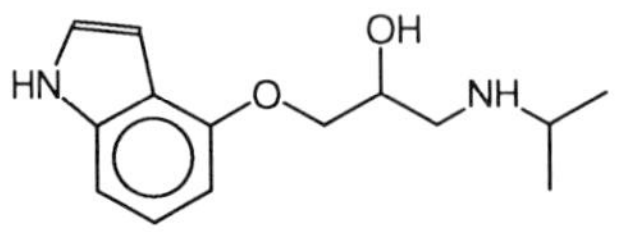

Pindolol

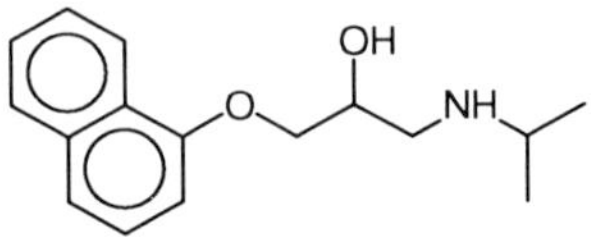

Propranolol

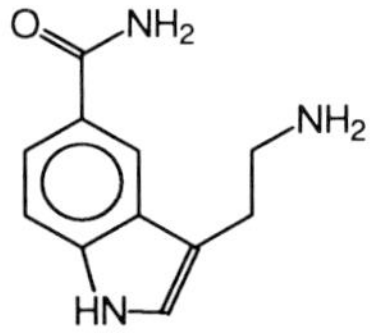

5-carboxamidotryptamine (5-CT)

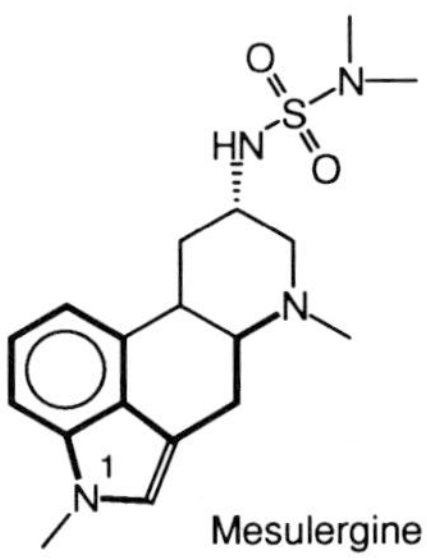

Mesulergine

Figure 6. Compounds used for identification of exogenous ligand binding sites. The bold lines in the structure of mesulergine represent the moiety that tryptamines and ergolines have in common.

Pindolol

									10										20							
1	2	3	4	5	6	7	8	9	0	1	2	3	4	5	6	7	8	9	0	1	2	3	4	5	6	7
+	+	+	+	+	+	+	–	–	–	–	–	–	–	–	–	–	–	–	–	–	–	–	–	–	–	–
N	N	N	N	N	N	N	T	T	T	T	A	A	A	V	V	L	L	L	L	T	T	Y	Y	Y	Y	Y

Propranolol

									10										20										30						
1	2	3	4	5	6	7	8	9	0	1	2	3	4	5	6	7	8	9	0	1	2	3	4	5	6	7	8	9	0	1	2	3	4	5	6
+	+	+	+	+	–	–	–	–	–	–	–	–	–	–	–	–	–	–	–	–	–	–	–	–	–	–	–	–	–	–	–	–	–	–	–
N	N	N	N	N	T	T	T	V	L	L	L	L	L	L	L	T	T	F	F	F	F	F	F	F	F	F	F	Y	Y	Y	Y	Y	I	I	I

Figure 7. Results of correlation analyses of propranolol and pindolol affinities for the sequences shown in *Table 3* for which binding data are available. The top lines indicate the sequence numbers in the same order as in *Table 3*. + and – indicate high and low affinity respectively. The bottom lines represent the amino acid at position 719 in the corresponding sequence.

Table 4. Affinity values[a] for 5-carboxamidotryptamine (5-CT), represented by a pseudo residue, and the corresponding receptors that were used in the analysis

Accession number	Pseudo residue[a]	Receptor type	Species
P08908	+	Serotonin 1A	Human
P28222	+	Serotonin 1B	Human
P28564	+	Serotonin 1B	Rat
P28334	+	Serotonin 1B	Mouse
P11614	+	Serotonin 1D	Dog
P28221	+	Serotonin 1D	Human
P28565	+	Serotonin 1D	Rat
P28566	–	Serotonin 1E	Human
P30939	–	Serotonin 1F	Human
Q02284	–	Serotonin 1F	Mouse
P30940	–	Serotonin 1F	Rat
P28223	–	Serotonin 2A	Human
P14842	–	Serotonin 2A	Rat
P30994	–	Serotonin 2B	Rat
P08909	–	Serotonin 2C	Rat
P30966	+	Serotonin 5A	Mouse
P35364	+	Serotonin 5A	Rat
P31387	+	Serotonin 5B	Mouse
P35365	+	Serotonin 5B	Rat
P31388	–	Serotonin 6	Rat
P34969	+	Serotonin 7	Human
P32304	+	Serotonin 7	Mouse
P32305	+	Serotonin 7	Rat

[a]Affinity values were taken from Boess and Martin (1993). Receptors with values $pK_i < 7.0$ are coded '–', receptors with $pK_i > 7$ are coded '+'. In addition, '+'-coded receptors displayed affinity 5-CT > 5-HT, while receptors coded '–' displayed affinity 5-CT < 5-HT.

```
                  10                  20
1 2 3 4 5 6 7 8 9 0 1 2 3 4 5 6 7 8 9 0 1 2 3
+ + + + + + + - - - - - - - - + + + + - + + +
P P P P P P P G N N N V V V P P P P P A P P P
```

Figure 8. Correlation analysis of the affinity of 5-CT for serotonin receptors shown in *Table 4*. The top line indicates the sequences in the same order as in *Table 4*. + and – indicate affinity differences as described in *Table 4*. The bottom line represents the amino acid at position 629 in the corresponding sequence.

Figure 8 shows that one single residue, 629, is correlated with high affinity for 5-CT. This proline is located two turns above (i.e. closer to the extracellular side) the 'PFF' motif, of which the second Phe (at position 622) has been shown to interact with agonists in the 5-HT_{2A} receptor and the β-adrenoceptor (Strader *et al.*, 1989; Choudhary *et al.*, 1993). This proline enables the formation of a hydrogen bond with the unsatisfied backbone C=O of residue 625. Thus, 5-CT may form a hydrogen bond with this backbone C=O, which is located only one helical turn above the putative binding site for agonists. In some models of the complexes of serotonin with 5-HT_{1A} and 5-HT_{2A} receptors, the 5-hydroxy group in serotonin is directed towards the backbone of helix VI (Kuipers *et al.*, 1994; Trumpp-Kallmeyer *et al.*, 1992). These models suggest that a hydrogen bond may be formed between the $-NH_2$ in 5-CT and the C=O of residue 625. Thus, the proline in position 629 may account for the high affinity of 5-CT for Pro-629-containing serotonin receptors, which makes this residue an interesting target for future mutagenesis studies.

Mesulergine binding site

Mesulergine has been classified as a 5-HT_{2C} receptor antagonist, but it also displays affinity for rat 5-HT_{2A} and all reported 5-HT_7 receptors (Boess and Martin, 1992). Its affinities for human 5-HT_{2A}, 5-HT_1 and 5-HT_6 receptors are much lower (Boess and Martin, 1993; Pazos *et al.*, 1985). We compared the sequences of the serotonin receptors of which the affinity for mesulergine was determined. We used pseudo-residue '+' for rat 5-HT_{2A} and all 5-HT_7 receptors, and '–' for human 5-HT_{2A} and all 5-HT_1 and 5-HT_6 receptors (*Table 5*). Unlike the cases of pindolol and propranolol, no single amino acid residue showed 100% correlation with affinity for mesulergine. Apparently, the affinity of mesulergine in the given data set depends on the presence or absence of more than just one residue. Analysing the correlation between the pseudo residue and two real residues at the same time, we were able to identify combinations of residues that might be involved in mesulergine affinity.

Figure 9 shows that the low affinity of mesulergine for 5-HT_1 and 5-HT_6 receptors may be explained by the presence of different residues in positions 127, 129, 223, 337 and 524, when compared with 5-HT_2 and 5-HT_7 receptors. The human 5-HT_{2A} receptor, which has low affinity for mesulergine, differs from other 5-HT_2 and 5-HT_7 receptors only in that residue 516 is a serine. Receptors with high affinity all contain an alanine at this position. The importance of this residue was confirmed by an Ala-516-Ser mutation in the rat 5-HT_{2A} receptor, which decreases the affinity for mesulergine as well as for a number of other N1-alkyl substituted ergolines and tryptamines (*Figure 6*, Johnson *et al.*, 1993). In contrast, the affinities of the un-substituted N1-H compounds increase as a result of the Ala-516-Ser mutation. Apparently, ergolines and tryptamines with a N1-alkyl substituent prefer an alanine at position 516, whereas the free N1-H compounds prefer a serine. These findings suggest a direct interaction between the N1-substituent and residue 516. Similarly, the mutation Ala-516-Thr in the rat 5-HT_{2A} receptor diminishes affinity for N1-alkyl substituted compounds (Johnson *et al.*, 1993). This explains the low affinity of mesulergine for the 5-HT_6 receptor which has a threonine at position 516. *Figure 9* shows that residues in other positions, such as 127, 129 and 223, may also contribute to the low affinity. These residues, as well as residue 524, are interesting candidates for future mutagenesis studies in 5-HT_1, 5-HT_5 and 5-HT_6 receptors.

Table 5. Affinity values for mesulergine, represented by a pseudo residue, and the corresponding receptor sequences which were used in the correlation analysis

Accession number	Pseudo residue[a]	Receptor type	Species
P14842	+	Serotonin 2A	Rat
P28335	+	Serotonin 2C	Human
P08909	+	Serotonin 2C	Rat
P30994	+	Serotonin 2B	Rat
P32304	+	Serotonin 7	Mouse
P32305	+	Serotonin 7	Rat
P34969	+	Serotonin 7	Human
P28223	–	Serotonin 2A	Human
P31388	–	Serotonin 6	Rat
P28564	–	Serotonin 1B	Dog
P28222	–	Serotonin 1B	Human
P28334	–	Serotonin 1B	Mouse
P28566	–	Serotonin 1E	Human
Q02284	–	Serotonin 1F	Mouse
P30939	–	Serotonin 1F	Human
P30940	–	Serotonin 1F	Rat
P30966	–	Serotonin 5A	Mouse
P35364	–	Serotonin 5A	Rat
P31387	–	Serotonin 5B	Mouse
P35365	–	Serotonin 5B	Rat

[a]Affinity values were taken from Seeman (1993), and Boess and Martin (1993). Values $pK_i > 7.0$ are coded '+', values $pK_i < 6.0$ are coded '–' .

Residue position	Sequence number 1	2	3	4	5	6	7	8	9	10	1	2	3	4	5	6	7	8	9	20
	+	+	+	+	+	+	+	–	–	–	–	–	–	–	–	–	–	–	–	–
127	I	I	I	I	I	I	I	I	A	T	T	T	T	T	T	T	F	F	F	F
129	G	G	G	G	G	G	G	G	A	S	S	S	L	I	I	I	W	W	W	W
223	A	A	A	A	A	A	A	A	S	T	T	T	T	T	T	T	S	S	S	S
337	S	S	S	S	S	S	S	S	S	A	A	A	A	A	A	A	A	A	A	A
516	A	A	A	A	A	A	A	S	T	A	A	A	A	A	A	A	A	A	A	A
524	M	M	M	M	M	M	M	M	I	L	L	L	I	I	I	I	V	V	V	V

Figure 9. Residue positions that show combined correlation with affinity for mesulergine in serotonin receptors.

8-OH-DPAT, DOI, flesinoxan and chlorpromazine

The compounds 8-OH-DPAT (2-(di-*n*-propylamino)-8-hydroxytetralin) and flesinoxan are highly selective full agonists for the 5-HT_{1A} receptor. Correlation analysis yields not only the residues important for binding, but all residues that distinguish the 5-HT_{1A} receptor from other amine receptors. Of the 27 residues that correlate well with high affinity for 8-OH-DPAT, 15 are in loop regions, not likely to be involved in agonist-binding. The same problem occurred with the 5-HT_2-selective agonist DOI (1-(2,5-dimethoxy-4-Br-phenyl)-2-aminopropane), yielding 17 residues which correlate with high affinity, of which five were in loop regions. Thus, for these highly selective compounds that only bind to one specific subclass of receptors, the signal from ligand binding is severely obscured by the signal created by other residue positions that distinguish this subclass from other classes and subclasses.

Another problem occurred with the antagonist chloropromazine, which is reported to display affinity for 5-$HT_{2A,2C}$, $dopamine_{2,3,4}$ and α_1-adrenoceptors (Boess and Martin, 1993; Seeman, 1993). A perfect correlation could not be observed when all the binding constants were included in one analysis. Apparently, the complete set is too complicated to be analysed by our method. Division of the set into subsets yielded too many hits for each analysis. The introduction of a scoring scheme that can correlate the pseudo-residue with three or four residues at the same time may solve this problem. On the other hand, it is doubtful whether the number of sequences (20–30 per compound) would, for such a scoring scheme, be sufficient for a clear signal. It seems likely that, in some cases, the receptor diversity is too large given the amount of available ligand binding data. This problem, however, is restricted to the use of ligand binding data and does not occur with functional classification for which all available sequences can be used.

It should also be kept in mind that the method contains a number of simplifications which may reduce its capability to extract correlations from the available data. For instance, affinities are coded '+' or '–' without any intermediate values. Also, only residue identities are used and physico-chemical similarities between different amino acid types are not taken into account. Thus a serine to threonine mutation is penalized to the same extent as a serine to tryptophan mutation. The nature of the data may also cause problems. When an antagonist is used as radioligand, a biphasic curve for the displacement with an agonist may be obtained, reflecting different affinities. If the number of receptors in the high-affinity state is low, it may be difficult to obtain a clear signal from high affinity agonist binding. As a result, reported agonist affinity values may be too low when determined with an antagonist as radioligand. Another problem is the fact that many receptor binding studies have involved mixtures of receptors or receptor subtypes, which makes it difficult to attribute the data to the individual proteins. Of course, one of the major assumptions is that a certain ligand with affinity for a number of receptors addresses an equivalent binding site on these receptors. Although our study shows that this is often a reasonable assumption, problems may be expected for compounds that do not conform to this criterion.

Despite these problems the method in its present form provides a tool for unbiased combination of sequence data with experimentally determined ligand affinities. In a number of cases, it is a useful tool for the identification of residues potentially involved in ligand binding. Residues for which a role in ligand binding has been experimentally determined were detected. In addition several suggestions for future mutagenesis experiments could be obtained.

4. Conclusions

We have presented a method of analysing sequence patterns in a multiple protein family. Pairwise comparisons of sequence positions can be used to search for functionally important residues without any prior knowledge. Residues can also be compared with properties of the proteins that are coded in so-called pseudo residues. These comparisons can be used to prove or disprove hypotheses, or to search for residues responsible for specific functional characteristics. For several ligand binding studies, our analyses led to a better understanding of the receptor models. In other cases, this correlation analysis has helped us circumvent the structure in the central protein research paradigm: sequence to structure to function.

Acknowledgements

The authors would like to thank Alfonso Valencia, Ulrike Goebel, Rob Hooft, Mike Singer, Georg Casari and Reinhard Schneider for many helpful discussions. FAPESP and IQCP provided financial support.

References

Bairoch A. Department Biochimie Medicinale, Centre Medicinal Universitaire, 1211, Geneva 4, Switzerland, and SWISS-PROT Protein sequence database, EMBL Data Library, D-69117 Heidelberg, Germany.

Boess FG, Martin IL. (1993) Molecular biology of 5-HT receptors. *Neuropharmacology*, **33**: 275–317.

Bowie JB, Lüthy R, Eisenberg D. (1991) A method to identify protein sequences that fold into a known three-dimensional structure. *Science*, **253**: 164–170.

Casari G, Sander C, Valencia A. (1995) A method to predict functional residues in proteins. *Struct. Biol.* **2**: 171–178.

Choudhary MS, Craigo S, Roth BL. (1993) A single point mutation (Phe340–Leu340) of a conserved phenylalanine abolishes 4-(125)iodo-(2,5)-dimethoxy)-phenylisopropylamine and (3H)-mesulergine but not (3H)ketanserin binding to 5-hydroxytryptamine 2 receptors. *Mol. Pharmacol.* **43**: 755–761.

Eck RV, Dayhoff MO. (1969) In: *Atlas of Protein Sequence and Structure* (Dayhoff MO, ed.) Washington DC: *Natl. Biomed. Res. Found.*, pp. 175–195.

Goebel U, Sander C, Schneider R, Valencia A. (1994) Correlated mutations and residue contacts in proteins. *Proteins,* **18**: 309–317.

Guan X, Peroutka SJ, Kobilka BK. (1992) Identification of a single amino acid residue responsible for the binding of a class of b-adrenergic receptor antagonists to 5-hydroxytryptamine1A receptors. *Mol. Pharmacol.* **41**: 695–698.

Henderson R, Baldwin JM, Ceska TA, Zemlin F, Beckmann E, Downing KH. (1990) Model of the structure of bacteriorhodopsin based on high-resolution cryo-microscopy. *J. Mol. Biol.* **212**: 899–929.

Hibert MF, Trumpp-Kallmeyer S, Bruinvels A, Hoflack J. (1991) Three-dimensional models of neurotransmitter G-binding protein-coupled receptors. *Mol. Pharmacol.* **40**: 8–15.

Johnson MP, Longcharich RJ, Baez M, Nelson DL. (1993) Species variations in transmembrane region V of the 5-hydroxytryptamine type 2A receptor alter the structure-activity relationship of certain ergolines and tryptamines. *Mol. Pharmacol.* **45**: 277–286.

Karnik SS, Khorana HG. (1990) Assembly of functional rhodopsin requires a disulfide bond between cysteine residues 110 and 187. *J. Biol. Chem.* **265**: 17520–17524.

Kolakowsky LF. (1994) GCRDb: A G-protein coupled receptor database. *Receptors and Channels*, **2**: 1–7.

Kuipers W, Wijngaarden I van, IJzerman AP. (1994) A model of the serotonin 5-HT1A receptor. Agonist and antagonist binding sites. *Drug Design Discovery.* **11**: 231–249.

NCI-FCRDC, Frederick Biomedical Supercomputing Center, PO Box B, Frederick, MD 21702-1201.

Oksenberg D, Marsters SA, O'Dowd BF, Jin H, Havlik S, Peroutka SJ, Ashkenazi A. (1992) A single amino-acid difference confers major pharmacological variation between human and rodent 5-HT1B receptors. *Nature*, **360**: 161–163.

Oliveira L, Paiva ACM, Sander C, Vriend G. (1994) A common step for signal transduction in G protein-couple receptors. *Trends Pharmacol. Sci.* **15**: 170–172.

Overington J, Donnelly D, Johnson MS, Sali A, Blundell TL. (1992) Environment-specific amino acid substitution tables: Tertiary templates and prediction of protein folds. *Prot. Sci.* **1**: 216–226.

Parker EM, Grisel DA, Iben LG, Shapiro RA. (1993) A single amino acid difference accounts for the pharmacological distinctions between the rat and human 5-hydroxytryptamine1B receptors. *J. Neurochem.* **60**: 380–383.

Pazos A, Hoyer D, Palacios JM. (1985) Mesulergine, a selective serotonin-2 ligand in the rat cortex, does not label these receptors in porcine and human cortex: evidence for species differences in brain serotonin-2-receptors. *Nature*, **360**: 161–163.

PIR, NBRF, Georgetown University Medical Center, 3900 Reservoir Road, NW Washington, DC.

Sakmar TP, Franke RR, Khorana HG. (1989) Glutamic acid-113 serves as the retinylidene Schiff base counterion. *Proc. Natl Acad. Sci. USA*, **86**: 8309–8313.

Sakmar TP, Franke RR, Khorana HG. (1991) The role of the retinylidene Schiff base counterion in rhodopsin in determining wavelength absorbance and Schiff base pKa. *Proc. Natl Acad. Sci. USA*, **88**: 3079–3084.

Sander C, Schneider R. (1991) Database of homology-derived protein structures and the structural meaning of sequence alignment. *Proteins*, **9**: 56–68.

Savarese TM, Fraser CM. (1992) *In vitro* mutagenesis and the search for structure-function relationships among G-protein-coupled receptors. *Biochem. J.* **283**: 1–19.

Scharf M. (1989) Thesis, University of Heidelberg.

Schertler FX, Villa C, Henderson R. (1993) Projection structure of rhodopsin. *Nature*, **362**: 770–772.

Seeman PMD. (1983) *Receptor Tables*, Vol. 2: *Drug dissociation constants for neuroreceptors and transporters.* Toronto: SZ Research.

Strader CD, Sigal IS, Dixon RAF. (1989) Structural basis of beta-adrenergic function. *FASEB J.* **3**: 1825–1832.

Suryanararayana S, Daunt DA, Zastrow M von, Kobilka BK. (1991) A point mutation in the seventh hydrophobic domain of the α2 adrenergic receptor increases its affinity for a family of β-receptor antagonists. *J. Biol. Chem.* **266**: 15488–15492.

Trumpp-Kallmeyer S, Hoflack J, Bruinvels A, Hibert M. (1992) Modeling of G-protein coupled receptors: Application to dopamine, adrenaline, serotonin, acetylcholine and mammalian opsin receptors. *J. Med. Chem.* **35**: 3448–3462.

Unger VM, Schertler GFX. (1995) Low resolution structure of bovine rhodopsin determined by electron cryomicroscopy. *Biophys. J.* **68**: 1776–1786.

Vriend G. (1990) WHAT IF: A molecular modeling and drug design program. *J. Mol. Graph.* **8**: 52–56.
Wang Z, Asenjo AB, Oprian DD. (1993) Identification of the Cl^--binding site in the human red and green color vision pigments. *Biochemistry*, **32**: 2125–2130.
Wess J, Gdula D, Brann MR. (1992) Site-directed mutagenesis of the muscarinic receptors: identification of a series of threonine and tyrosine residues involved in agonist but not antagonist binding. *EMBO J.* **10**: 3728–3734.
Zhukovsky E, Oprian DD. (1989) Effect of carboxylic acid chains on the absorbance maximum of visual pigments. *Science*, **246**: 928–930.

Chapter 3

Modelling α-helical transmembrane domains

D. Donnelly, N. Bhogal, C.R. Hurrell and J.B.C. Findlay

1. Introduction

The necessity for the reliable prediction of protein structure from sequence is nowhere more marked than for integral membrane proteins where the rate of structure determination is hampered by difficulties in obtaining sufficient quantities of material for structural studies. In this chapter we outline an approach towards obtaining structural information from sequence alignments by describing the construction of a general model for the α-helical transmembrane domain of the rhodopsin-like G protein-coupled receptors (GPCRs).

2. Predicting structure from sequence

A general three-dimensional model of the α-helical transmembrane domain of the rhodopsin-like GPCRs was derived from data obtained from the analysis of a sequence alignment of 59 aminergic GPCRs. The method uses a Fourier transform-based algorithm embodied in a group of computer programs called PERSCAN (Donnelly *et al.*, 1993, 1994a) and relies upon the different properties of lipid-accessible and protein-buried residues. In particular, lipid accessible positions tend to mutate at a faster rate than those that are protein-buried, and they are usually uncharged. Hence, the residue substitution patterns at each position in a sequence alignment can provide information about the extent to which that position is exposed to or hidden from the lipid phase of the bilayer. For a transmembrane segment, the exposure of each residue (at position j) can be represented as a value S_j and together these values form the substitution profile S. Since other profiles can also be used to estimate exposure (e.g. variability (V) or hydrophobicity (H)), we will use the general term U to represent a profile consisting of elements U_j.

Hence, a property profile U is calculated so that a measure of exposure U_j is assigned at each position j in a sequence or sequence alignment over a window size N. The moment M can be calculated as

$$M = \left\{ \left[\sum_{j=1}^{N} U_j \sin(j\omega) \right]^2 + \left[\sum_{j=1}^{N} U_j \cos(j\omega) \right]^2 \right\}^{1/2} \qquad (1)$$

where ω is the angle between adjacent side chains when the sequence is considered as a regular structure and viewed down an axis defined by the C_α atoms. If the values of U_j represent the hydrophobicities (H_j) of the residues then M is the hydrophobic moment (μ, Eisenberg *et al.*, 1984). When calculating the periodicity in the values of U_j, the Fourier transform power spectrum is calculated by

$$P(\omega) = [x]^2 + [y]^2 \tag{2}$$

where

$$x = \sum_{j=1}^{N} U_j^n \sin(j\omega) \tag{3}$$

$$y = \sum_{j=1}^{N} U_j^n \cos(j\omega) \tag{4}$$

and where

$$U_j^n = U_j - \bar{U} \quad (j = 1, 2, \ldots N) \tag{5}$$

$\bar{U}$ is the average value of U_j over the window. U^n is therefore a normalized version of the profile U adjusted so that $\sum_{j=1}^{N} U_j^n = 0$ and hence $\bar{U}^n = 0$. The new profile U^n consists of elements U_j^n in which the periodicity in internal/external residues is predicted by the periodicity in positive/negative values (or negative/positive values) of U_j^n. This results in cleaner power spectra (since $P(\omega) = 0$ when $\omega = 0$) so that the α-periodicity index AP can be calculated as

$$\mathrm{AP} = \left[(1/30) \int_{90^\circ}^{120^\circ} P(\omega)\, d\omega\right] \Big/ \left[(1/180) \int_{0^\circ}^{180^\circ} P(\omega)\, d\omega\right] \tag{6}$$

AP is analogous to ψ used in Komiya *et al.* (1988) and to the amphipathic index AI used in Cornette *et al.* (1987) (although the precise boundaries of the helical regions of the power spectrum differ in the latter). AP is a ratio of the extent of the periodicity in the helical region of the spectrum to that over the whole spectrum. Komiya *et al.* (1988) suggest that a value of AP greater than 2 indicates that the helical periodicity is significant. This indicates the presence of a helix within the window of sequence used but it is difficult to predict the precise start and finish of the helix. More recently, we have used a slightly altered definition of AP for the prediction of transmembrane α-helices (Donnelly *et al.*, 1994a).

$$\mathrm{AP}^{\mathrm{n}} = \left[(1/11) \int_{95^\circ}^{106^\circ} P(\omega)\, d\omega\right] \Big/ \left[(1/180) \int_{0^\circ}^{180^\circ} P(\omega)\, d\omega\right] \tag{7}$$

This definition results in a greater value of AP^{n} for power spectra with sharper, more narrow peaks above α-helical region of the spectrum where $\omega = 100°$. Narrow maxima are more typical of longer regions of periodic sequence and this should be the case for transmembrane helices.

The direction of the internal face of a predicted helix can be estimated from the direction of the moment $\sqrt{P(\omega)}$ when $\omega = 100°$. This is the moment produced by the profile U^n when the sequences form an ideal α helix. θ is the angle describing the direction of the moment relative to the first residue ($j = 1$). θ can be calculated by

$$\gamma = \arccos[y/\sqrt{P(\omega)}] \quad (8)$$

where γ is greater than 0° and less than 180°.

$$\theta = \begin{cases} \gamma & \text{if } x > 0 \\ 360 - \gamma & \text{if } x < 0 \end{cases} \quad (9)$$

The point at which the lipid-accessible face contacts the more polar environment of the phosphate head groups and surrounding aqueous environment can be estimated by the point at which charged residues appear on this side of the helix. Charged residues should not be present on the lipid facing side of a transmembrane helix.

The analogous calculations can be carried out to calculate the direction of the hydrophobic and conserved face of the helix so that the results of the three analyses can be compared. This allows us to predict whether the helix is in an aqueous or lipid environment since the conserved and hydrophobic faces should be equivalent for the former but opposite for the latter. This can be useful for predicting the point at which a transmembrane helix protrudes from the lipid region of the bilayer into the more polar regions at the bilayer boundaries since the direction of the hydrophobic face should change by approximately 180° relative to the direction of the conserved/internal face (Donnelly and Cogdell, 1993; Donnelly *et al.*, 1994a).

The Fourier transform analysis of the amino acid substitution patterns observed in sequence alignments may therefore provide a means of predicting and orientating α-helices. In order to test this, the method was applied to a sequence alignment of three members of the bacteriorhodopsin family (Donnelly *et al.*, 1993). The estimated orientation of each of the helices correlated well with the structure and suggests that the method can be used for prediction. The method has also been applied to the prediction of α-helices in water-soluble proteins (Donnelly *et al.*, 1994b).

3. Models of GPCRs

The PERSCAN suite of programs was used to analyse an alignment of 59 GPCRs that bind biogenic amine ligands (Donnelly *et al.*, 1994a). Each of the seven transmembrane segments was found to possess strong helical periodicity and hence the orientation of each helix relative to the remainder of the protein could be determined.

The point at which the fifth transmembrane helix protrudes into the cytoplasm and the point at which the seventh transmembrane helix protrudes on the extracellular side of the bilayer are detected by the relative direction of the hydrophobic and buried faces as described above. There appears to be a small non-helical region dividing the main transmembrane helices from the shorter protruding helical regions. For helix 7, the non-helical region includes the residues Asn-312 and Trp-313 (residues and numbers correpond to the human β_2-adrenergic receptor: hβ_2-AR). One consequence of this prediction is that the position equivalent to the retinal binding lysine residue in rhodopsin (Lys-296; equivalent to Tyr-316 in hβ_2-AR) is very close to the extracellular side of the bilayer. This is in contrast to the retinal binding lysine in bacteriorhodopsin (Lys-216) which is buried in the centre of the bilayer.

Based on these data, a three-dimensional model of the human β_2-adrenergic receptor was constructed. The arrangement of the helices was based on that observed in the projection map of bovine rhodopsin (Schertler *et al.*, 1993) using a connectivity analogous to that observed in bacteriorhodopsin (Baldwin, 1993; Henderson et al., 1990). Adrenaline can be docked into the receptor model in agreement with the published data (i.e. Asp-113, Ser-204, Ser-207 and Phe-290 interact with the ligand; Strader *et al.*, 1987, 1989ab). Additional interactions can be predicted from the model, for example Asn-293 (on helix 6) can interact with the β-hydroxyl group of the ligand while Ser-319 (helix 7) can interact with Asp-113 (helix 3). Additional GPCR models (e.g. the human neurokinin, NK2, receptor and human rhodopsin) were constructed from the original β_2-adrenergic receptor model by replacing the side chains within the computer graphics program INSIGHTII (Biosym Technologies, San Diego, CA, USA). The models were then subjected to energy minimization in order to remove the many unfavourable contacts.

4. The peptide binding site of the NK2 receptor

Since ligand binding to both the β_2-adrenergic receptor and to rhodopsin is believed to involve residues on the transmembrane helices, we hypothesized that similar regions of the human NK2 receptor would form part of the NKA binding site although extracellular regions may also be required to bind this larger ligand. A model of the transmembrane regions of the human NK2 receptor was used to identify potential NKA binding residues on transmembrane helices 2, 3, 5, 6 and 7. Subsequent site-directed mutagenesis experiments and ligand binding analyses have identified several residues that may form part of the NKA binding site (Bhogal *et al.*, 1994; Hurrell *et al.*, 1995). All of these mutants (*Table 1*), with the exception of His-198-Leu and His-198-Ala, maintain wild type binding affinity for the non-peptidic antagonist ^{3}H-SR48,968 but show reduced or abolished binding of the physiological peptidic ligand NKA. *Figure 1a* shows a ribbon model of the human NK2 receptor with the putative NKA binding residues displayed with dot surfaces. The putative binding site for NKA lies between helices 2, 3, 5, 6 and 7 and is in close agreement with the ligand binding sites of the human β_2-adrenergic receptor (*Figure 1b*) and human rhodopsin (*Figure 1c*). Indeed, many of the putative NKA binding residues are at topologically equivalent positions to the ligand binding residues identified in these receptors.

Table 1. Ligand binding data for the wild type and mutant human NK2 receptors

Receptor	IC_{50} (nM, 0.1 nM^{125}NKA/NKA)	Kd (nM, ^{3}H-SR48,968)
Wild type	1.1±0.1	2.7±0.3
Asn-86-Thr	NB	7.1±0.3
Gln-109-His	NB	2.5±0.5
His-198-Ala	NB	NB
His-198-Leu	NB	NB
Ile-202-Val	NB	4.0±0.3
Gly-273-Pro	NB	10.7±0.1
Gly-273-Thr	NB	2.0±0.5
Met-297-Leu	8.5±2.4	1.6±0.3

NB, no detectable binding of radio-ligand.

The maintenance of antagonist binding shows that these mutant receptors are expressed and are structurally intact. However, it should be noted that agonists are believed to bind with high affinity only to the active conformation of the receptor (R*) which can couple to G protein. Antagonists on the other hand, are thought to bind with high affinity to both the active (R*) and inactive (R) forms of the receptor (e.g. see Rosenkilde *et al.*, 1994 and references within). It is therefore possible that the effect of the mutations is to alter the ability of the receptor to undergo these conformational changes so that the equilibrium between R and R* for the mutant receptors is pushed towards the inactive R form. Such mutations may be distant from the NKA binding site despite the observation that NKA binding is apparently reduced or abolished. Hence, site-directed mutagenesis may be a dangerous tool for probing agonist–receptor interactions and more direct approaches involving cross-linking of receptor and ligand may be necessary in order to obtain more concrete data. Such data will improve our models and lead to a greater understanding of these important receptors.

(a)

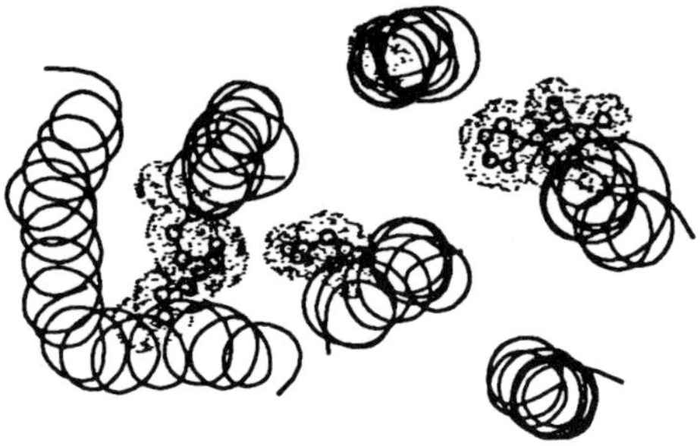

(b)

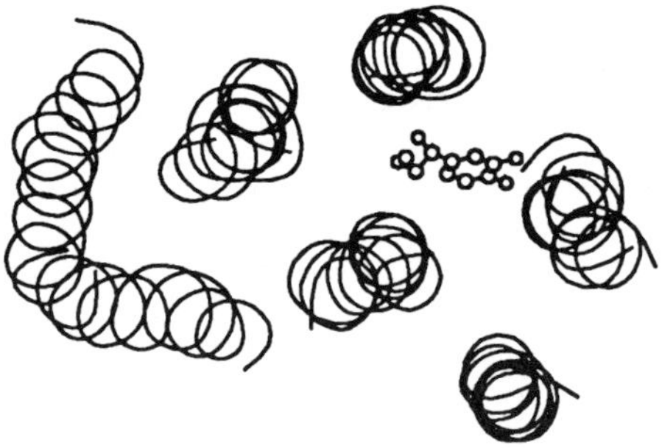

(c)

Figure 1. A single stranded ribbon representation of three GPCR models. (a) a model of the human NK2 receptor with the putative NKA binding residues high-lighted with dot surfaces. This binding site correlates closely to that observed for the adrenaline binding site in the β-adrenergic receptor (b) and to the chromophore binding site in rhodopsin (c). Ligands are shown in ball and stick representation.

Acknowledgements

The authors are grateful to Professor T.L. Blundell and Dr J.P. Overington for their valuable contribution to the molecular modelling work described in this chapter.

References

Baldwin JM. (1993) The probable arrangement of the helices in G protein-coupled receptors *EMBO J.* **12**: 1693–1703.

Bhogal N, Donnelly D, Findlay JBC. (1994) The ligand binding site of the neurokinin 2 receptor. *J. Biol. Chem.* **269**: 27269–27274.

Cornette JL, Cease KB, Margalit H, Spouge JL, Berzofsky JA, DeLisi C. (1987) Hydrophobicity scales and computational techniques for detecting amphipathic structures in proteins. *J. Mol. Biol.* **195**: 659–685.

Donnelly D, Cogdell RJ. (1993) Predicting the point at which transmembrane helices protrude from the bilayer: a model of the antenna complexes from photosynthetic bacteria. *Protein Eng.* **6**: 629–635.

Donnelly D, Overington JP, Ruffle SV, Nugent JHA, Blundell TL. (1993) Modeling α-helical transmembrane domains: the calculation and use of substitution tables for lipid-facing residues. *Protein Sci.* **2**: 55–70.

Donnelly D, Findlay JBC, Blundell TL. (1994a) The evolution and structure of aminergic G protein-coupled receptors. *Receptors Channels,* **2**: 61–78.

Donnelly D, Overington JP, Blundell TL. (1994b) The prediction and orientation of α-helices from sequence alignments: the combined use of environment-dependent subtitution tables, Fourier transform methods and helix capping rules. *Protein Eng.* **7**: 645–653.

Eisenberg D, Weiss RM, Terwilliger TC. (1984) The hydrophobic moment detects periodicity in protein hydrophobicity. *Proc. Natl Acad. Sci. USA* **81**: 140–144.

Henderson R, Baldwin JM, Ceska TA, Zemlin F, Beckmann E, Downing KH. (1990) Model for the structure of bacteriorhodopsin based on high-resolution electron cryomicroscopy. *J. Mol. Biol.* **213**: 899–929.

Hurrell, C.R., Donnelly, D., Findlay, J.B.C. (1995) Design and construction of a mutant neurokinin-2 receptor with convenient peptide cleavage sites. *Protein Peptide Lett.* **2**: 321–326.

Komiya H, Yeates TO, Rees DC, Allen JP, Feher G. (1988) Structure of the reaction center from Rhodobacter sphaeroides R-26: Symmetry relations and sequence comparisons between different species. *Proc. Natl. Acad. Sci. USA,* **85**: 9012–9016.

Rosenkilde MM, Cahir M, Gether U, Hjorth SA, Schwartz TW. (1994) Mutations along transmembrane segment II of the NK-1 receptor affect substance P competition with non-peptide antagonists but not substance P binding. *J. Biol. Chem.* **269**: 28160–28164.

Schertler GFX, Villa C, Henderson R. (1993) Projection structure of rhodopsin. *Nature* **362**: 770–772.

Strader CD, Sigal IS, Register RB, Candelore MR, Rands E, Dixon RAF. (1987) Identification of residues required for ligand binding to the β-adrenergic receptor. *Proc. Natl. Acad. Sci. USA*, 4384–4388.

Strader CD, Candelore MR, Hill WS, Sigal IS, Dixon RAF. (1989a) Identification of two serine residues involved in agonist activation of the beta-adrenergic receptor. *J. Biol. Chem.* **264**: 13572–13578.

Strader CD, Sigal IS, Dixon RAF. (1989b) Structural basis of β-adrenergic receptor function. *FASEB J.* **3**: 1825–1832.

Chapter 4

α-Helix bundles and ion channels

R. Sankararamakrishnan and M.S.P. Sansom

1. Introduction

A parallel bundle of transmembrane (TM) α-helices surrounding a central pore is a structural motif present in several classes of ion channel (Oiki *et al.*, 1990; Sansom, 1991). This motif occurs in the nicotinic acetylcholine receptor (nAChR), the central pore of which is formed by a bundle of five M2 helices (Sansom, 1992, 1993a; Stroud *et al.*, 1990; Unwin, 1989; 1993). Parallel helix bundles are also generated by a number of channel-forming peptides (CFPs), which self-assemble within lipid bilayers to form transbilayer pores. Such CFPs include alamethicin (Sansom, 1993bc) and *Staphylococcus aureus* δ-toxin (Mellor *et al.*, 1988; Sansom 1991). In order to understand the molecular basis of ion channel formation it is important to develop models of bundles of parallel TM helices.

Several investigators have used molecular modelling to study packing of α-helices. Energy minimization studies have been carried out on dimers of simple hydrophobic helices by Chou *et al.* (1984) and by Furois-Corbin and Pullman (1986a). The latter authors extended this approach to bundles of helices (Furois-Corbin and Pullman, 1986b; 1987). More recently, simulated annealing via restrained molecular dynamics simulations (SA/MD) has been developed as a molecular modelling tool, and has been applied to the GCN4 leucine zipper (Nilges and Brünger, 1991), and glycophorin TM helix (Treutlein *et al.*, 1992) dimers. This method generates an ensemble of structures which satisfy distance restraints chosen to embody biochemical data. An advantage of SA/MD is that it enables exploration of a wider range of conformational space than is feasible using energy minimization or constant temperature molecular dynamics simulations.

In the current study, SA/MD is used to generate models of pores formed by pentameric bundles of: (a) simple hydrophobic TM helices; (b) a simple amphipathic TM helix; and (c) a chick α7 nAChR M2 helix. Bundles are analysed with respect to their geometric properties, including the dimensions of the central pore. A preliminary model for the nAChR pore is discussed, in which the constituent M2 helices tilt outwards from the central axis in order to accommodate the channel-blocking local anaesthetic derivative QX-222.

2. Methods

Molecular dynamics (MD) simulations were carried out using Xplor V3.1 (Brünger, 1992) and the CHARMM PARAM19 (Brooks *et al.*, 1983) parameter set. Only those hydrogen atoms attached to polar groups were represented explicitly, apolar groups being represented using extended atoms. QX-222 atomic partial charges were calculated using MOPAC. Display and examination of models was carried out using Quanta V3.2 (Molecular Simulations), and diagrams of structures drawn using Molscript (Kraulis, 1991). MD simulations were performed on Convex C2 and DEC 3000 400 computers. All other calculations were on a Silicon Graphics Indigo R3000 workstation.

The SA/MD method used is similar to that of Nilges and Brünger (1991). It has been described with respect to single TM helices (Kerr and Sansom, 1993; Sankararamakrishnan and Sansom, 1994) and TM-helix bundles (Kerr *et al.*, 1994ab), and so only a brief description follows. The starting point is a Cα template corresponding to a pentameric bundle of exactly parallel α-helices. Remaining backbone and side chain atoms are built automatically by simulated annealing, during which the Cα atoms remain fixed. Cα restraints are then gradually released whilst electrostatic interactions are introduced. Inter-helix distance restraints ensure overall pseudo-symmetry of the bundles during a final (5 ps) burst of constant temperature (300 K) molecular dynamics. The overall procedure is repeated to generate ensembles of 20 structures for each Cα template.

To analyse the resultant structures, helix crossing angles (Ω) were determined as described by Chothia *et al.*, (1981). Inter-helix separations between adjacent helices of a bundle (D) were determined by calculating the distance between a pair of virtual atoms defined, in each helix, by the geometric centre of the Cα atoms of the helix. Averages of Ω and D were determined across all adjacent helix pairs within an ensemble. Helix–helix interaction energies (ΔE) were calculated as:

$$\Delta E = E(\text{bundle}) - \Sigma\, E(\text{constituent helices})$$

Pore dimensions were estimated using HOLE (Smart *et al.*, 1993; Kerr *et al.*, 1994b). Pore radii are displayed as a function of the distance moved parallel to the axis of rotational symmetry of the Cα template (z), thus generating pore radius profiles.

3. Results

Simple hydrophobic helices

Ensembles of 20 structures for pentameric, parallel bundles of Ala-20 and Leu-20 were generated in order to evaluate the method, and to investigate packing of simple hydrophobic helices. As can be seen from *Table 1*, the termini of the helices were blocked, in order to mimic the preceding and following peptide bonds within a protein. *Table 2* summarizes the geometric and energetic properties of the helix bundles. Of interest are the crossing angles (Ω) between adjacent helices within the bundles. The initial value was Ω = 0°, i.e. exactly parallel helices. In both cases the final value of Ω is positive, indicating that helix packing within the bundles corresponds to class 3–4 'ridges-in-grooves' helix packing as described by Chothia (1984). A positive value for the crossing angle corresponds to a left-handed coiled-coil structure for the bundle as a whole.

Helix bundles within each ensemble were ranked in order of their five-fold rotational symmetry. Figure 1(*a*, *b*) shows the most symmetrical Ala-20 bundle in 'ribbons' format. Although the structure has approximate five-fold symmetry, the use of inter-helix distance restraints during SA/MD permits small deviations from exact symmetry of the bundle. The left-handed coiled-coil structure is evident. The distribution of Ω values across the entire Ala-20 ensemble is given in *Figure 1*(*c*). Most crossing angles fall within the range +10 to +17.5°. The mean inter-helix separation, ⟨*D*⟩ = 7.8 Å, indicates close packing of the constituent helices of the bundle. Examining the structures in more detail reveals that the Cβ atoms of the side chains interdigitate to yield 'ridges-in-grooves" packing.

The geometrical properties of the Leu-20 ensemble (*Table* 2) were similar to those for Ala-20, the major difference being an increase in inter-helix separation (*D*) from 7.9 to 9.7 Å. The distribution of crossing angles for Leu-20 was almost identical to that for Ala-20. Furthermore, the most symmetrical Leu-20 bundle had a coiled coil pitch (*P* = 283 Å) close to that of the corresponding Ala-20 bundle (*P* = 234 Å). The

Table 1. Transmembrane helix sequences

Helix	Sequence
Ala-20	Ac-A-A-A-A-A-A-A-A-A-A-A-A-A-A-A-A-A-A-A-A-NH_2
Leu-20	Ac-L-L-L-L-L-L-L-L-L-L-L-L-L-L-L-L-L-L-L-L-NH_2
M2C	AC-L-L-I-**S**-L-A-I-**S**-V-L-**L**-**S**-L-T-V-F-L-L-L-I-NH_2
M2α7	Ac-**E**-K-I-**S**-L-G-I-**T**-V-L-**L**-**S**-L-T-V-F-M-L-L-V-A-E-NH_2

(Channel-lining residues at positions 1′, 4′, 8′, 11′ and 12′ of M2 are in bold).

Table 2. Ensemble statistics

Model	⟨Ω⟩ (degrees)	⟨*D*⟩ (Å)	⟨ΔE^{VDW}⟩ (kcal mol^{-1})
Ala-20	+13 (2)	7.9 (0.2)	−121 (14)
Leu-20	+13 (2)	9.7 (0.2)	−200 (6)
M2c	+8 (5)	9.5 (0.7)	−155 (8)
M2α7P	+13 (3)	9.5 (0.1)	−191 (8)
M2α7T	+6 (7)	9.2 (0.2)	−118 (10)

Helix crossing angles (Ω), inter-helix separations (*D*), and the van der Waals component of the helix–helix interaction energies (⟨ΔE^{VDW}⟩) are given as a mean (SD) for each ensemble of 20 structures.

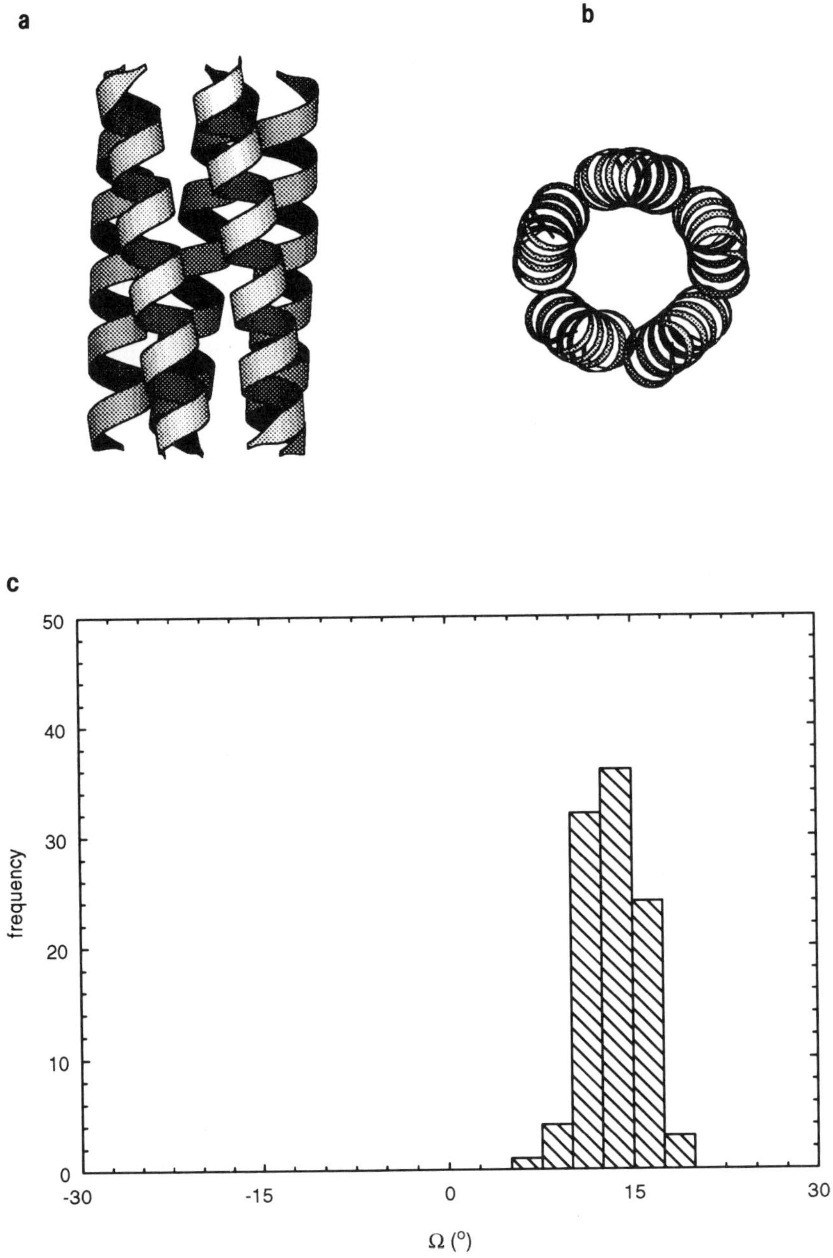

Figure 1. The most symmetrical Ala-20 helix bundle, displayed in 'ribbons' fashion. (a) the bundle viewed down a perpendicular to the pore *(z)* axis with the N-termini of the helices at the top of the diagram. (b) A view along *z* from the N-termini towards the C-termini. (c) The distribution of helix crossing angles (Ω) for the Ala-20 ensemble.

van der Waals component of the helix–helix interaction energy (ΔE^{VDW}) was larger for Leu-20 than for Ala-20 bundles (ca. –40 and –20 kcal mol^{-1}/helix respectively), reflecting the larger side chains in the interfaces of the former bundles.

The dimensions of the central pores formed by the Ala-20 and Leu-20 bundles were estimated using HOLE (Smart *et al.*, 1993), which determines the pore radius (R) as a function of the distance travelled parallel to the pore, (i.e. the z, axis). This is achieved by allowing the centre of the pore, for each z value, to vary within the xy plane so as to maximize the pore radius, thus defining an irregular 'trajectory" for the centre of the pore. The results presented (*Table 3* and *Figure 2*) are for the most symmetrical member of each ensemble. However, the same analysis applied to the least symmetrical members of the same ensembles did not yield significantly different results.

HOLE calculations provide information on pore-lining residues of the helix bundles. Both the Ala-20 and Leu-20 bundles contain a central pore from z ca. 0 Å (the N-terminal mouth) to z ca. 30 Å (the C-terminal mouth). There is a periodic variation in the pore radius. In both models the closest contact atoms at constrictions of the pores are the Cβ atoms of residues 3′, 6′, 10′, 13′ and 17′. Such periodicity clearly reflects that of the constituent helices, and indicates positions at which one might expect to find polar side chains in the pore-lining helices of channel proteins. Note that despite the greater helix–helix separation of the Leu-20 bundle, the mean and minimum pore radii are less than for Ala-20. The minimum radius (R_{MIN} = 1.3 Å) of the Leu-20 pore corresponds to, for example, the crystal radius of a K^+ ion.

Table 3. Pore properties

Model	⟨R⟩ (Å)	R_{MIN} (Å)	Closest contacts
Ala-20	2.2	1.9	Cβ: A–3
		1.9	Cβ: A–6
		2.0	Cβ: A–10
		1.8	Cβ: A–13
		1.7	Cβ: A–17
Leu-20	1.8	1.5	Cδ: L–3
		1.4	Cδ: L–6
		1.3	Cδ: L–10
		1.4	Cδ: L–13
		1.3	Cδ: L–17
M2c	2.2	1.0	Cδ: L–11
		1.3	Cδ: L–19
M2α7P	2.7	1.4	Cδ: L–11
		0.9	Cδ: L–19
M2α7T	3.6	1.2	Oε: E–1

For the most symmetric bundle from each ensemble the mean pore radius (⟨R⟩) and the radii at constrictions of the pore (RMIN) are given, along with the atoms forming the closest contacts at the constrictions.

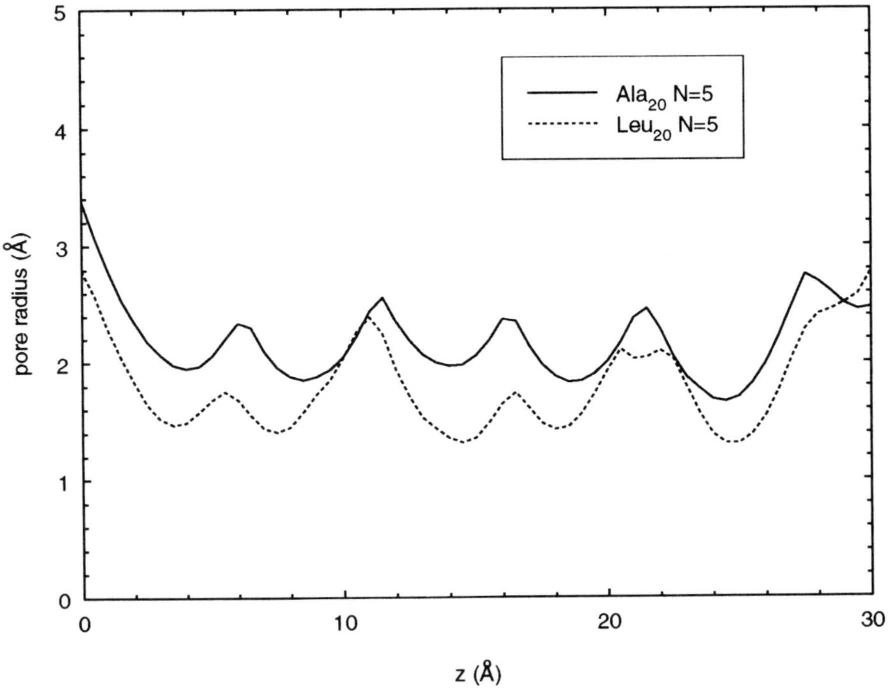

Figure 2. Pore radius profiles calculated using HOLE for the most symmetrical Ala-20 (Ala20) (solid line) and Leu-20 (Leu20) (dotted line) helix bundles. The N-termini of the helices are at *z* ca. 0 Å and the C-termini at *z* ca. 30 Å.

A simple amphipathic helix

Ion channels are generally formed by bundles of amphipathic TM helices. The sequence of the amphipathic helix M2c (*Table 1*) is derived from the sequence of the pore-lining M2 helices of nicotinic acetylcholine receptors (nAChR; Lester, 1992; Stroud *et al.*, 1990; Unwin, 1993). Site-directed mutagenesis (Charnet *et al.*, 1990; Leonard *et al.*, 1988; Villarroel *et al.*, 1991), chemical labelling (Bertrand *et al.*, 1993; Changeux *et al.*, 1992; Giraudat *et al.*, 1987; Hucho *et al.*, 1986) and molecular modelling (Sansom, 1992) studies suggest that serine and/or threonine side chains at positions 4′, 8′ and 12′ of nAChR M2 helices interact with ions as they pass through the channel. M2c is a consensus sequence derived by comparison of M2 helices from different nAChRs. However, unlike nAChR M2 sequences (e.g. M2α7, *Table 1*), M2c does not contain ionizable side chains at the termini of the helix. These were excluded in order to focus on properties of hydroxyl-containing groups within the transbilayer region of M2.

An ensemble of 20 pentameric bundles of M2c helices was generated. The Cα template was constructed such that the centre of the hydrophilic face of each helix, defined by the Cα atoms of S8′, was directed towards the centre of the bundle. Thus, the three serine side chains formed the lining of the resultant pore, as can be seen in *Figure 3(b)*. The geometric and energetic properties of the ensemble are summarized in *Table 2*, and the properties of the pore formed by the most symmetrical bundle are given in *Table 3*.

M2c helices formed bundles corresponding to a left-handed coiled coil (*Figure 3(a)*) in a similar fashion to Ala-20 and Leu-20. However, the mean crossing angle for the M2c ensemble was somewhat lower than those of the two hydrophobic helix ensembles (*Table 2*). Furthermore, the distribution of crossing angles is broader for the M2c ensemble than for Ala-20 and Leu-20, indicating a greater heterogeneity of helix packing within the M2c ensemble. Inter-helix separations within the M2c bundles do not differ significantly from those within Leu-20 bundles. The ΔE^{VDW} value for the M2c ensemble is intermediate between those of Ala-20 and Leu-20.

The pore radius profile for the M2c bundle (*Figure 3(c)*) reveals a less regular pattern of constrictions than for Ala-20 or Leu-20. The constriction of the pore near its N-terminal mouth (at z ca. 3 Å) is formed, in part, by the hydroxyl groups of S4′ side chains. Mutagenesis studies suggest that residue 4′ of nAChR M2 may correspond to the narrowest region of the open channel of nAChR (Villaroel *et al.*, 1991). The central constriction of the M2c bundle (at ca. $z = 15$ Å) is formed by the side chains of residue L11′. The radius at this point falls to 1.0 Å (i.e. less than the ionic radius of K^+), effectively occluding the pore. Residue 11′ of nAChR M2 is a highly conserved leucine which has been implicated in occluding the channel when the latter is in a closed (or desensitized) conformation (Revah *et al.*, 1991; Unwin, 1993). Thus, the M2c model has several features reminiscent of the nAChR channel.

Nicotinic acetylcholine receptor

As discussed above, the pore domain of nAChR is formed by a pentameric bundle of M2 helices. We have used SA/MD to model the structure of this domain of the intact channel protein. An additional element of experimental evidence provides valuable constraints on possible models of the structure of the nAChR pore when the channel is in an open conformation. Certain hydrophobic cationic molecules (non-competitive open channel blockers) bind within the pore. Such open channel blockers include local anaesthetics (e.g. procaine, lidocaine, benzocaine) and compounds derived from local anaesthetics (e.g. QX-222) (F*igure 4(a)*). Site-directed mutagenesis studies demonstrate that QX-222 interacts with serine residues at positions 8′ and 12′ of M2 (Leonard *et al.*, 1988; Charnet *et al.*, 1990), and with the leu-cine at position 11′ (Revah *et al.*, 1991). Potential energy calculations using GRID (Goodford, 1985) suggest that the carbonyl oxygen of QX-222 may form hydrogen-bonds with the side chain hydroxyl groups of residues 8′ and 12′ of M2 (Busby and Sansom, unpublished results). Thus, models of the nAChR pore should accommodate QX-222 in the vicinity of residues 8′, 11′ and 12′.

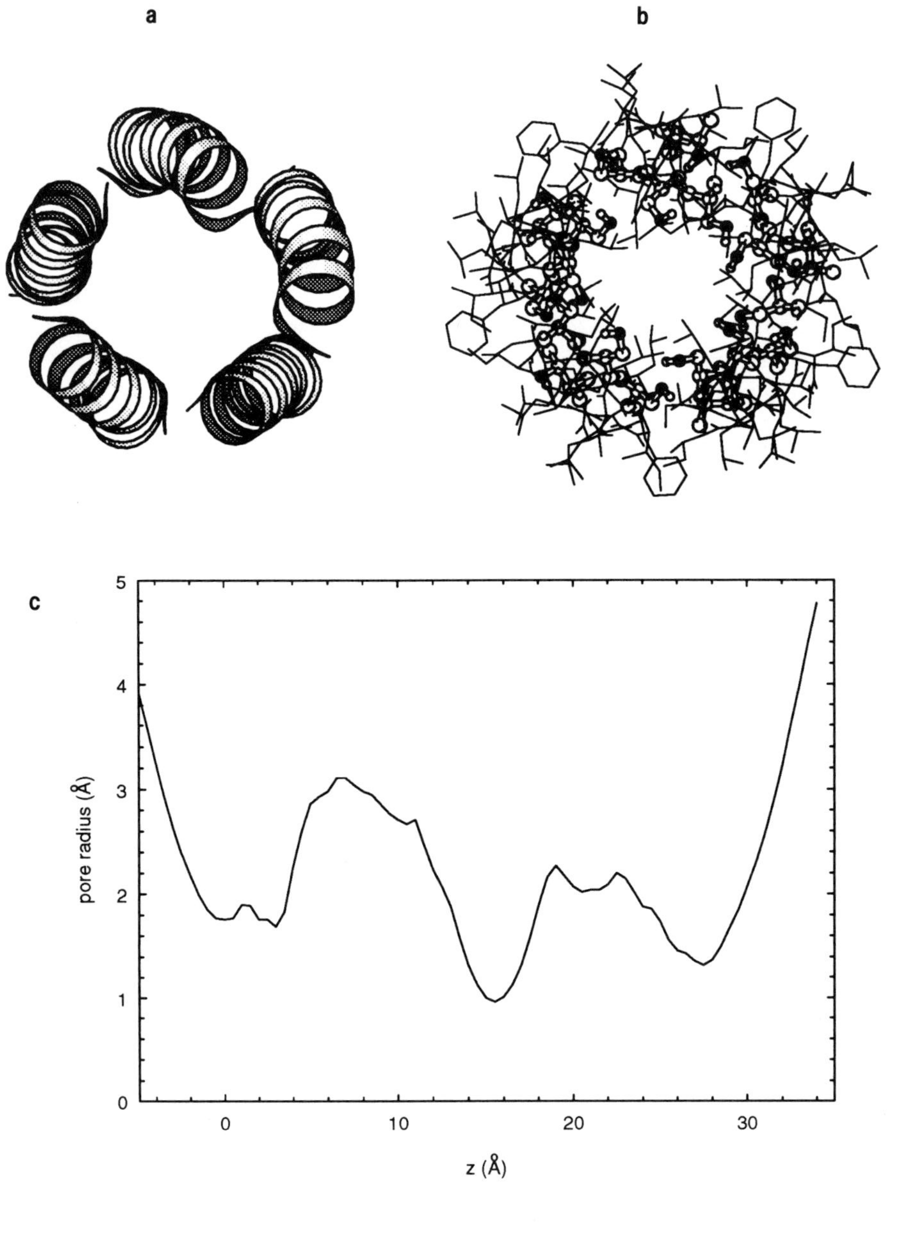

Figure 3. The most symmetrical bundle structure for amphipathic helix M2c. (a) The bundle shown in ribbon format, (b) all atoms are shown, with serine and threonine residues highlighted using 'ball-and-stick' format. The view is down the *z* axis from the N-termini towards the C-termini, (c) the corresponding pore radius profile, the N-termini of the helices being at *z* ca. 0 Å.

a

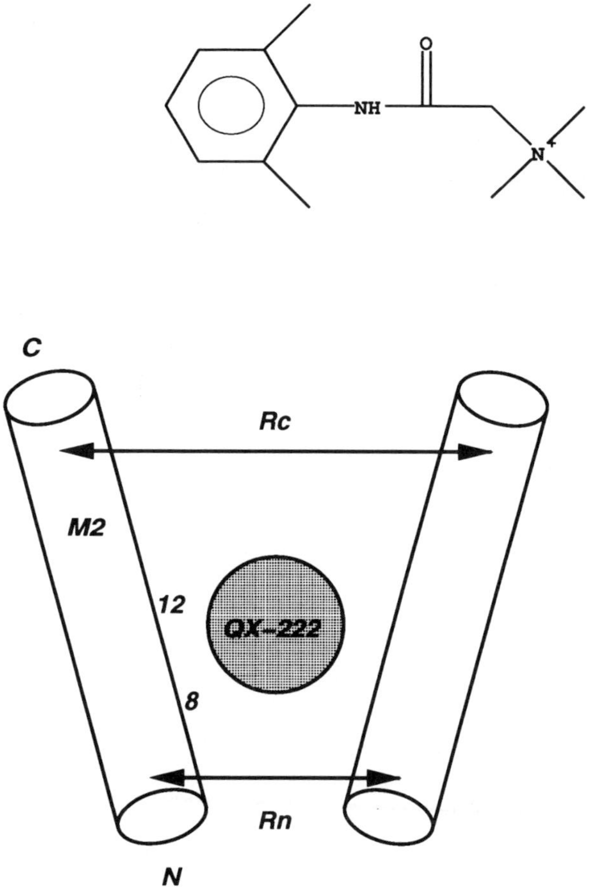

Figure 4. QX-222 binding to the nAChR channel. (a) Shows the covalent structure of QX-222. (b) A schematic diagram of how the M2 helices need to be tilted in order to accommodate QX-222 within the pore.

For our initial studies we have modelled chick α7 neuronal nAChR pores (*Table 1*). This nicotinic receptor subunit is able to form homo-pentameric (i.e. $\alpha 7_5$) channels, and has been the subject of extensive mutagenesis studies (Bertrand *et al.*, 1993). Two ensembles of M2α7 models were developed (a) M2α7P, in which the helices in the Cα template were exactly parallel, and (b) M2α7T, in which the helices of the Cα template were tilted away from the *z* axis so as to increase the C-terminal radius of the pore. Thus, model M2α7P was constructed without including information on QX-222 binding, whereas M2α7T incorporated QX-222-derived distance restraints during generation of the model.

M2α7P bundle

This ensemble was generated in the same manner as for M2c. The initial orientation of the M2α7 helices within the Cα template was such that the Cα of T8′ was directed towards the centre of the pore. From *Table 2* and *Figure 5*(*a*) it can be seen that the resultant bundles are left-handed coiled-coils, with crossing angles and inter-helix separations comparable to those of Leu-20. The ΔE^{VDW} value is also close to that of the Leu-20 bundle. Thus in M2α7P the helices pack with efficient interdigitation of the interfacial side chains. However, as is evident in *Figure 5*(*a*), the M2α7P helices are distorted towards their C-termini. Examination of the (ϕ, ψ) values suggests that residues 15′ to 17′ are distorted away from α-helical conformations. Preliminary studies suggest that this may occur in order to relieve the mutual repulsion of the C-terminal glutamate (E22′) residues.

As can be seen from *Figure 5*(*b*) the central pore is lined by the hydroxyl groups of residues 4′, 8′ and 12′. The pore radius profile for M2α7P (*Figure 5*(*c*)) is similar to that for M2c. M2α7P has a narrower constriction than M2c at the N-terminus, corresponding to the presence of the E1′ side chains in the M2α7 helices. HOLE analysis (*Table 3*) shows that the pore has a mean radius of less than 1.5Å in two places, corresponding to the rings of side chains of residues L11′ and L19′. Both constrictions decrease the mean radius of the pore to approximately 1.3 Å. Whereas small side chain movements might be expected to enable a K^+ ion, for example, to 'squeeze' through, it is evident that QX-222 (maximum van der Waals radius ca. 5 Å, see below) cannot be accommodated within the M2α7P pore. This is particularly so given that QX-222 enters the pore from the C-terminal (i.e. extracellular) mouth, which is occluded by the ring of L19′ side chains. Thus, modifications of the nAChR model are required in order to match the experimental data on the open channel conformation.

M2α7T bundle

The nature of the M2α7T model is indicated schematically in *Figure 4*(*b*). A simple search of QX-222 conformations was carried out using Quanta. Low energy conformations thus generated were used to calculate the maximum radius of a cylinder which would accommodate QX-222, yielding a van der Waals contact radius of 5.0 Å. The M2 helices were tilted so as to accommodate QX-222 in the vicinity of residue L11′, i.e. between residues 8′ and 12′ (see above). Treating the helices as cylinders of radius 4.7 Å, this corresponds to helix axis to pore axis distances of (a) R_N = 8 Å at the N-termini of the helices (b) $R_{L11'}$ = 9.7 Å in the vicinity of L11′; and (c) R_C = 11.3 Å at the C-termini. A grid search of Eulerian angles defining the M2 helix orientations was used to generate Cα templates consistent with (a) the constraints on *R*, and (b) minimization of the sum of Cα–Cα distances for T8′ residues on opposite sides of the pore. The latter constraint ensured

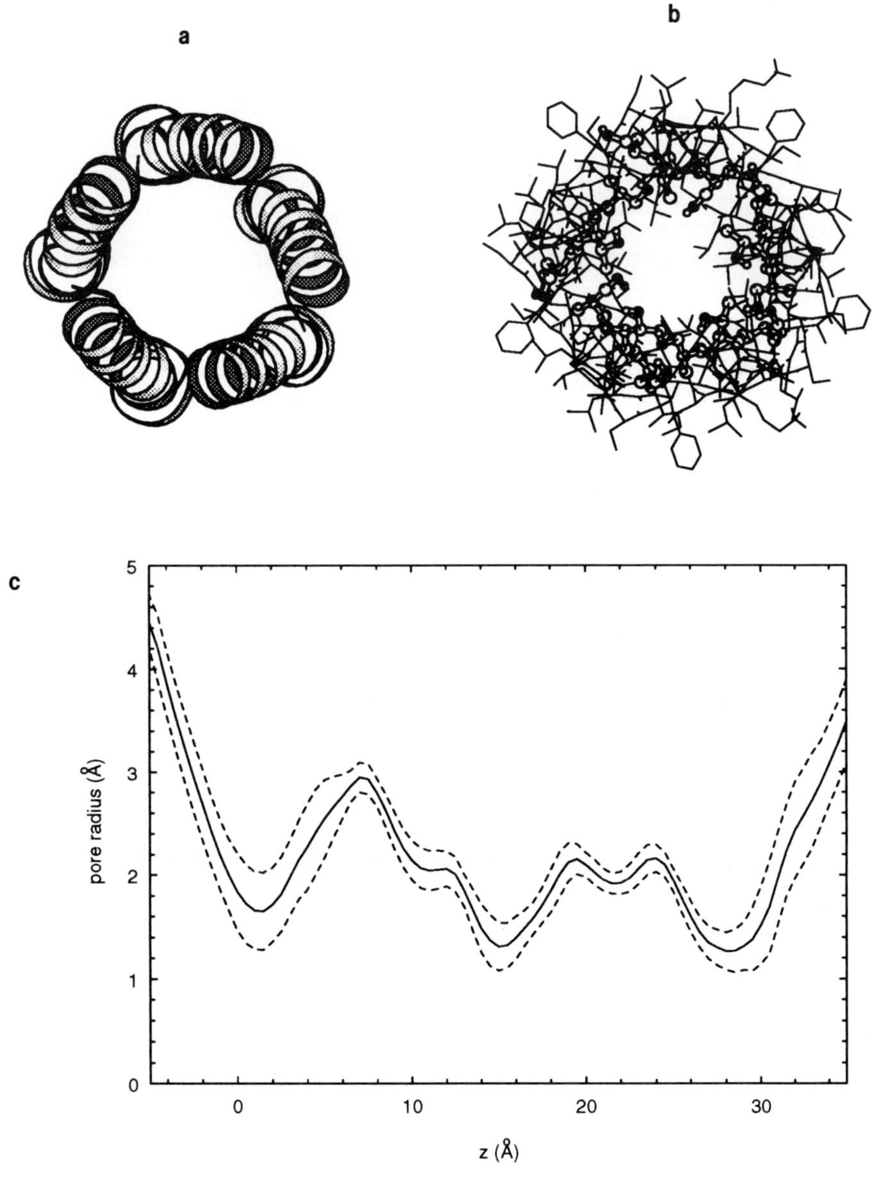

Figure 5. The most symmetrical M2α7P bundle. (a) The bundle shown in ribbon format, (b) all atoms are shown with residues 4′, 8′ and 12′ highlighted using 'ball-and-stick' format. The view is down *z* from the N-termini towards the C-termini. (c) The corresponding pore radius profile, the N-termini of the helices being at *z* ca. 0 Å. The solid line indicates the mean pore radius, averaged across the ensemble of 20 structures, ±1 SD (dotted lines).

that the hydrophilic face of M2, defined by Cα:T8′, was directed towards the centre of the pore. One of the resultant Cα template bundles was used for SA/MD. This bundle had a helix tilt angle (from the central z axis) of approximately 6°. This should be compared with the value of 7° of Furois-Corbin and Pullman (1989a,b), who used chlorpromazine binding within the pore to guide their modelling of nAChR. During the final MD burst of SA/MD, inter-helix distance restraints were employed in order to maintain the helix tilt.

A 'ribbons' diagram of the most symmetrical member of the resultant ensemble is shown in *Figure 6(a)*, viewed from the N-termini of the helices. The bundle still forms a left-handed coiled-coil, even though the helices are tilted relative to the central axis, as confirmed by analysis of helix crossing angles (*Table 2*). Thus 'ridges-in grooves' packing of the closely packed N-termini of the helices occurs. The interactions between the helices are, as expected, weaker for M2α7T than for M2α7P. Thus, ΔE^{VDW} for M2α7T is about the same as for Ala-20, i.e. less than for the other helix bundles. Examination of the structure in *Figure 6(b)* suggests that the aromatic side chains of F16′ may contribute to the helix–helix interfaces. As before, the hydroxyl groups of residues 4′, 8′ and 12′ line the central pore. Distortion of the helices in the vicinity of residue 16′ is absent from M2α7T.

The pore radius profile for M2α7T is given in *Figure 6(c)*. The constrictions of the pore in the vicinity of L11′ and L19′ seen in M2α7P are absent from M2α7T. Indeed, in the vicinity of L11′ the pore appears to be sufficiently wide to accommodate QX-222. (Note that the van der Waals radius of 5 Å for the QX-222 enveloping cylinder represents the maximum radius of the LA molecule, i.e. corresponding to the two methyl groups attached to the aromatic ring.) The N-terminal mouth of the pore is constricted by the side chains of E1′ residues. However, these side chains could move in response to local electrostatic fields generated by permeant cations.

Refinement of the M2α7T model

In order to evaluate the interactions of the M2α7T model with open channel blockers, a QX-222 interaction energy profile was evaluated. A QX-222 molecule was translated parallel to the pore (i.e. z) axis. For each position on z, the centre of QX-222 molecule was placed at the corresponding position on the HOLE trajectory, and an energy minimization performed to optimize QX-222/channel interactions. During the minimization the centre of the QX-222 molecule was restrained to the xy-plane and the Cα coordinates of the bundle subjected to harmonic positional restraints. All other atoms were free to move. At the end of the minimization the change in potential energy of the system was calculated.

A QX-222/M2α7T bundle complex as defined by the minimum in the QX-222 interaction energy profile is shown in *Figure 7*. This places the local anaesthetic

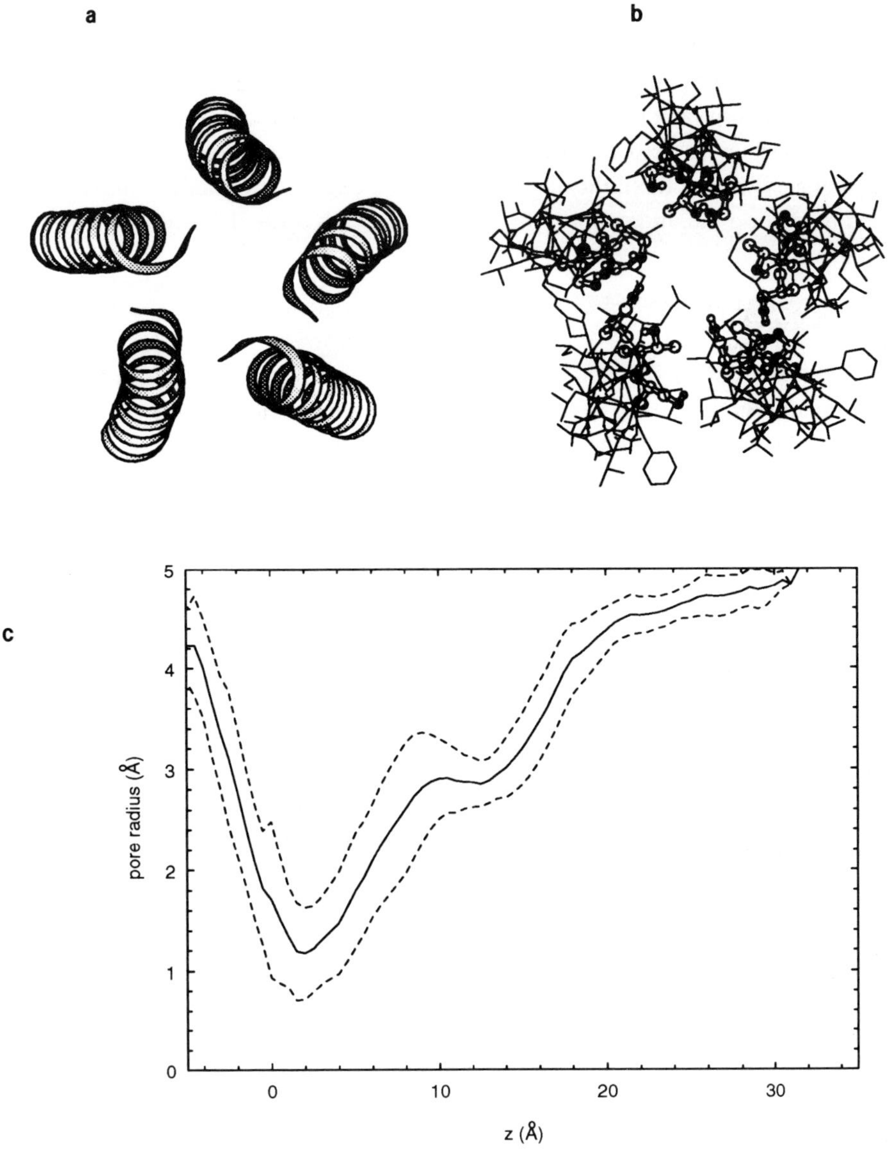

Figure 6. The most symmetrical M2α7T bundle. (a) The bundle shown in ribbon format, (b) All atoms are shown. Other details are as for *Figure 4*. (c) The mean pore radius profile (solid line) ±SD (dotted line).

molecule in the vicinity of side chains 1′, 4′ and 8′. This is possibly too far towards the N-terminal mouth of the pore, as experimental observations suggest that the side chains of residue 12′ form contacts with QX-222. Further refinement of the pore model has been undertaken in order to optimize the fit to the experimental data, yielding model M2α7T2. In model M2α7T2, the degree of tilt of the helices is midway between that of M2α7P (tilt angle 0°) and M2α7T (tilt angle 6°). In this refined model (M2α7T2), the QX-222 binding site, as defined by the minimum in the corresponding interaction profile, is in the vicinity of side chain 8′ and 12′ (i.e. in agreement with experimental observations). These results emphasize the importance of taking flexibility of ligand and side chain conformations into account when modelling the interactions of pore domains with open channel blockers.

4. Conclusions

These studies confirm that SA/MD is a useful technique for automatic generation of helix bundle models using experimentally derived distance restraints. The method minimizes user intervention and thus reduces bias in helix packing and side chain conformations. When applied to the parallel helix bundle motif SA/MD generates conformational features which are consistent with known structural properties of such proteins, and which were not explicitly included in the Cα template. For example, the X-ray structure of p-LI (a synthetic GCN4-derived peptide) reveals a left-handed coiled-coil of four parallel helices, with a pitch of 205 Å (Harbury *et al.*, 1993). This is similar to the packing geometry observed in the SA/MD-derived hydrophobic TM helix bundles described above.

A recent extension of the method, illustrated by the M2α7T and QX-222 study, uses approximate distance restraints to generate a ligand binding site within a helix bundle. This approach may be generally applicable to integral membrane proteins.

Two features of α-helix bundle structures of relevance to ion channels have been analysed (a) pore radius profiles, and (b) pore-lining side chains. The radii of pores formed by bundles of simple hydrophobic TM helices may be compared with those of other ion channels. Pentameric bundles of Ala-20 and of Leu-20 have mean pore radii ca. 2.2 Å and 1.8 Å, and minimum radii of ca. 1.7 Å and 1.3 Å, respectively. Thus either pore is wide enough to permit passage of, for example, an unhydrated K^+ ion (radius 1.3 Å). This may be compared with the $\leftrightarrow\beta^{6.3}$ model of the gramicidin A (GA) channel (analysed by Smart *et al.*, 1993) which has a mean pore radius of ca. 1.5 Å, and a minimum radius of 1.2 Å. Thus, pentameric bundles of parallel α-helices possess wider pores than GA, even before taking account of the selective presence of small polar side chains in the pore lining.

Pore radius profiles for both the simple hydrophobic TM helix bundles showed α-helix-related periodicity, corresponding to 'rings' of side chains lining the pore. As has been discussed by Bertrand *et al.* (1993) there is considerable experimental'evidence for such 'rings' of side chains in the lining of the nAChR pore. If such 'rings' contain small side chains the radius of the pore is increased. This is

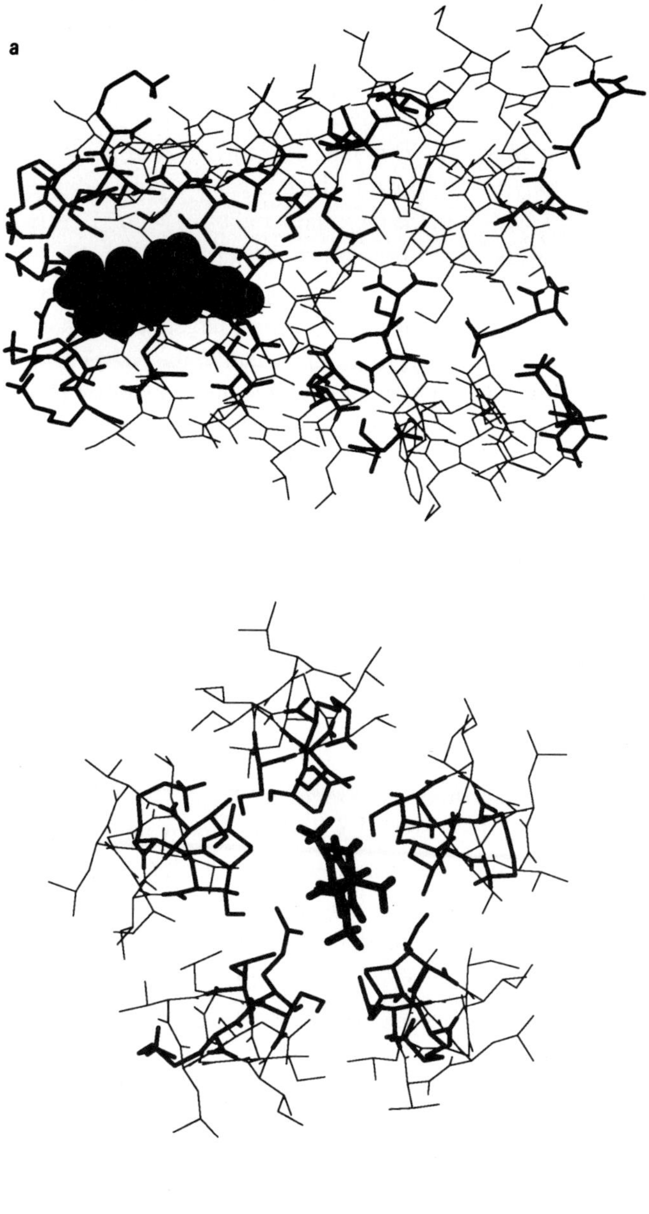

Figure 7. Two views of the M2α7T/QX-222 complex, with QX-222 positioned in the minimum in the corresponding interaction energy profile. (a) The M2 helix bundle is shown perpendicular to the pore axis, with the N-termini of the helices on the left hand side of the diagram. The foremost M2 helix is omitted for clarity. The QX-222 molecule is shown as a silhouette. (b) The bundle is viewed down the pore axis, and only those residues (1′ to 11′ which interact with QX-222 are indicated. In both diagrams the following code for residues is used: thick lines = QX-222; bold lines = polar residues; thin lines = apolar residues.

found in the amphipathic M2c helix bundle, for which the mean radius is ca. 2.2 Å compared with 1.8 Å for Leu-20, albeit with a pronounced constriction in the vicinity of L11′.

The same method has been applied to generate a 'first generation' model for the pore domain of a nAChR, i.e. M2α7T2. This has a mean radius of ca. 2.5 Å, in apparent conflict with electrophysiological data which suggest a pore radius of ca. 3.3 Å for nAChR (Hille, 1992). However, the mean pore radius of M2α7T2 is in part determined by a constriction at the N-terminal mouth of the pore. It is likely that changes in the conformation of E1′ side chains in response to local electrostatic fields may modulate this. The interaction of M2α7T2 with large, permeant cations is being investigated to test this possibility. More importantly, the M2α7T2 model appears to accommodate local anaesthetics and related open channel blockers within the centre of the pore, between residues 8′ and 12′.

M2α7T2 is a model of the pore domain of the nAChR when the latter is in an 'open' conformation. Current studies are concerned with refining this model, and using it to develop an atomic resolution model of the interaction of LAs within the nAChR pore. Future studies will extend this approach to modelling the pores of related receptor-gated channels, (e.g. those of GABA receptors), and to modelling their interactions with channel blocking drug molecules (Galzi and Changeux, 1994).

Acknowledgements

This work was supported by a grant from the Wellcome Trust. Our thanks to our colleagues J. Breed, I.D. Kerr and H.S. Son, and to Dr O. Smart for providing us with his program HOLE.

References

Bertrand D, Galzi JL, Devillers-Thiéry A, Bertrand S, Changeux JP. (1993) Stratification of the channel domain in neurotransmitter receptors. *Curr. Opin. Cell Biol.* **5**: 688–693.

Brooks BR, Bruccoleri RE, Olafson BD, States DJ, Swaminathan S, Karplus M. (1983) CHARMM: a program for macromolecular energy, minimization, and dynamics calculations. *J. Comp. Chem.* **4**: 187–217.

Brünger AT. (1992) X-PLOR, Version 3.1, *A system for X-ray crystallography and NMR.* New Haven, CT: Yale University Press.

Changeux JP, Galzi JI, Devillers-Thiéry A, Bertrand D. (1992) The functional architecture of the acetylcholine nicotinic receptor explored by affinity labelling and site-directed mutagenesis. *Quart. Rev. Biophys.* **25**: 395–432.

Charnet P, Labarca C, Leonard RJ, Vogelaar NJ, Czyzyk L, Gouin A, Davidson N, Lester HA. (1990) An open-channel blocker interacts with adjacent turns of α-helices in the nicotinic acetylcholine receptor. *Neuron,* **2**: 87–95.

Chothia C. (1984) Principles that determine the structure of proteins. *Ann. Rev. Biochem.* **53**: 537–572.

Chothia C, Levitt M, Richardson D. (1981) Helix to helix packing in proteins. *J. Mol. Biol.* **145**: 215–250.

Chou KC, Némethy G, Scheraga HA. (1984) Energetic approach to the packing of α-helices. 2. General treatment of nonequivalent and nonregular helices. *J. Am. Chem. Soc.* **106**: 3161–3170.

Furois-Corbin S, Pullman A. (1986a) Theoretical study of the packing of a-helices by energy minimization: effect of the length of the helices on the packing energy and on the optimal configuration of a pair. *Chem. Phys. Lett.* **123**: 305–310.

Furois-Corbin S, Pullman A. (1986b) Theoretical study of the packing of α-helices of poly(L-ala-nine) into transmembrane bundles. Possible significance for ion transfer. *Biochim. Biophys. Acta,* **860**: 165-177.

Furois-Corbin S, Pullman A. (1987) Theoretical study of the packing of α-helices into possible transmembrane bundles: sequences including alanines, leucines and serines. *Biochim. Biophys. Acta,* **902**: 31-45.

Furois-Corbin S, Pullman A. (1989a) A possible model for the inner wall of the acetylcholine receptor channel. *Biochim. Biophys. Acta,* **984**: 339-350

Furois-Corbin S, Pullman A. (1989b) Energy profiles in the acetylcholine receptor (AChR) channel. *FEBS Lett.* **252**: 63-68.

Galzi JL, Changeux JP. (1994) Neurotransmitter-gated ion channels as unconventional allosteric proteins. *Curr. Opin. Struct. Biol.* **4**: 554–565.

Goodford PJ. (1985) A computational procedure for determining energetically favorable binding sites on biologically important macromolecules. *J. Med. Chem.* **28**: 849–857.

Giraudat J, Dennis M, Heidmann T Haumont PY, Lederer F, Changeux JP. (1987) Structure of the high-affinity binding site for noncompetitive blockers of the acetylcholine receptor: [^{3}H] chlorpromazine labels homologous residues in the β and δ chains. *Biochemistry,* **26**: 2410–2418.

Harbury PB, Zhang T, Kim PS, Alber T. (1993) A switch between two-, three-, and four-stranded coiled coils in GCN4 leucine zipper mutants. *Science,* **262**: 1401–1407.

Hille B. (1992) *Ionic Channels of Excitable Membranes*, 2nd edn. Sunderland, MA: Sinauer Associates Inc.

Hucho F, Oberthür, W, Lottspeich, F. (1986) The ion channel of the nicotinic acetylcholine receptor is formed by the homologous helices M II of the receptor subunits. *FEBS Lett.* **205**: 137–142.

Kerr, ID, Sansom, MSP. (1993) Hydrophilic surface maps of α-helical channel-forming peptides. *Eur. Biophys. J.* **22**: 269–277.

Kerr, ID, Sankararamakrishnan, R, Sansom, MSP. (1994a) Simplified models of the pore domain of the nicotinic acetylcholine receptor. *Biochem. Soc. Trans.* **22**: 158S.

Kerr, ID, Sankararamakrishnan, R, Smart, OS, Sansom, MSP. (1994b) Parallel helix bundles and ion channels:- molecular modelling via simulated annealing and restrained molecular dynamics. *Biophys. J.* **67**: 1501–1515.

Kraulis, PJ. (1991) MOLSCRIPT: a program to produce both detailed and schematic plots of protein structures. *J. Appl. Cryst.* **24**: 946–950.

Leonard, RJ, Labarca, CG, Charnet, P, Davidson, N, Lester, HA. (1988) Evidence that the M2 membrane-spanning region lines the ion channel pore of the nicotinic receptor. *Science,* **242**: 1578–1581.

Lester, H. (1992) The permeation pathway of neurotransmitter-gated ion channels. *Ann. Rev. Biophys. Biomol. Struct.* **21**: 267–292.

Mellor, IR, Thomas, DH, Sansom, MSP. (1988) Properties of ion channels formed by Staphylococcus aureus δ-toxin. *Biochim. Biophys. Acta,* **942**: 280–294.

Nilges M, Brünger AT. (1991) Automated modelling of coiled coils: application to the GCN4 dimerization region. *Prot. Eng.* **4**: 649–659.

Oiki S, Madison V, Montal M. (1990) Bundles of amphipathic transmembrane α-helices as a structural motif for ion-conducting channel proteins: studies on sodium channels and acetylcholine receptors. *Proteins: Struct. Funct. Genet.* **8**: 226–236.

Revah F, Bertrand D, Galzi JL, Devillers-Thiery A, Mulle C, Hussy N, Bertrand S, Ballivet M, Changeux JP. (1991) Mutations in the channel domain alter the desensitization of a neuronal nicotinic receptor. *Nature,* **353**: 846–849.

Sankararamakrishnan R, Sansom MSP. (1995) Structural features of isolated M2 helices of nicotinic receptors: simulated annealing via molecular dynamics studies. *Biophys. Chem.* **55**: 215–230.

Sansom MSP. (1991) The biophysics of peptide models of ion channels. *Prog. Biophys. Mol. Biol.* **55**: 139–236.

Sansom MSP. (1992) An investigation of the role of serine and threonine side chains in ion channel proteins. *Eur. Biophys. J.* **21**: 281–298.

Sansom MSP. (1993a) Acetylcholine receptor: Peering down a pore. *Curr. Biol.* **3**: 239–241.

Sansom MSP. (1993b) Ala-methicin and related peptaibols – model ion channels. *Eur. Biophys. J.* **22**: 105–124.

Sansom MSP. (1993c) Structure and function of channel-forming peptaibols. *Quart. Rev. Biophys.* **26**: 365–421.

Smart OS, Goodfellow JM, Wallace BA. (1993) The pore dimensions of gramicidin A. *Biophys. J.* **65**: 2455–2460.

Stroud RM, McCarthy MP, Shuster M. (1990) Nicotinic acetylcholine receptor superfamily of ligand gated ion channels. *Biochemistry,* **50**: 1107–11023.

Treutlein HR, Lemmon MA, Engelmann DM, Brünger AT. (1992) The glycophorin A transmembrane domain dimer: sequence-specific propensity for a right-handed supercoil of helices. *Biochemistry,* **31**: 12726–12733.

Unwin N. (1989) The structure of ion channels in membranes of excitable cells. *Neuron,* **3**: 665–676.

Unwin N. (1993) Nicotinic acetylcholine receptor at 9 Å resolution. *J. Mol. Biol.* **230**: 1101–1124.

Villarroel A, Herlitze S, Koenen M, Sakmann B. (1991) Location of a threonine residue in the α-subunit M2 transmembrane segment that determines the ion flow through the acetylcholine receptor channel. *Proc. R. Soc. Lond. B,* **243**: 69–74.

Chapter 5

Making and breaking models of G protein-coupled receptors

Pamela Thomas

1. Introduction

The family of G protein-coupled receptors (GPCRs) is a large and important class of membrane bound proteins. More than 600 sequences are known for members of this family of proteins, but as yet our knowledge of their structure is extremely limited. The lack of experimentally determined structures has lead to the description of increasing numbers of GPCR models (see, for example, Cronet *et al.*, 1993; Findlay and Pappin, 1986; Hibert *et al.*, 1991; Lewell, 1992; Livingstone, *et al.*, 1992; Maloney-Huss and Lybrand, 1992).

This chapter will look at some of the experimental data obtained for this family of proteins and some of the information gleaned from sequence analysis. It will go on to consider how this information can be used to derive suitable templates for making models of GPCRs and some of the potential pitfalls in obtaining the final model.

2. What do experimental studies tell us about the structure of G protein-coupled receptors?

This section does not attempt to review all the data available from investigations into the structure of GPCRs. It concentrates on those data which have been used in this work to derive a suitable template for GPCR models.

Direct determination of the structure of a GPCR

The structures of several thousand proteins have been determined. However, very few of these structures are of membrane bound proteins. This reflects the difficulty in crystallizing membrane bound proteins and, in the case of the GPCR family, the difficulty in obtaining sufficient quantities of protein for crystallization trials. The only members of the family of GPCRs to have proved amenable to direct structural determination to date are the rhodopsins. This is due to the fact that these proteins are naturally expressed in large amounts in the disk membranes of the rod and cone cells in the retina.

The preparation and analysis of two-dimensional crystals of both bovine and frog rhodopsins have been reported. The projection map of frog rhodopsin dimers at 22 Å resolution (Corless *et al*, 1982) shows that the dimer has a sigmoidal shape and dimensions of 20–25 Å by 70–80 Å. The projection map of bovine rhodopsin has been calculated at 9 Å resolution (Schertler *et al.*, 1993). The overall shape is ellipsoidal and there are peaks in the projection map which can be interpreted as seven membrane spanning helices. Four of the helices lie perpendicular to the plane of the lipid bilayer, appearing as four discrete peaks in the projection map. The remaining three are tilted with respect to the plane of the membrane and appear as an 'arc-shaped feature' in the projection map. Estimates of the length and width of bovine rhodopsin monomer based on the published map and unit cell dimensions are consistent with the reported dimensions of the frog rhodopsin dimer.

The structure of bacteriorhodopsin

Bacteriorhodopsin (bR) is a light-activated proton pump found in the membranes of the purple bacterium, *Halobacter halobium.* In these membranes the protein forms two-dimensional crystals. This feature of the protein has been exploited to determine a low resolution structure of bR by cryo-electron microscopy (Henderson *et al.*, 1990). The resolution of the resulting structure is 3.5 Å in the plane of the membrane and 10 Å perpendicular to the plane. Bacteriorhodopsin has seven membrane spanning helices and a chromophore, retinal, which is attached to a lysine residue on the seventh helix via a Schiff's base. The structure and the packing of the membrane spanning helices in bR is shown in *Figure 1.*

The existence of seven membrane-spanning helices and the attachment of retinal to a lysine residue on the seventh membrane-spanning helix, is suggestive of a relationship between the structures of bR and eukaryotic rhodopsins (and hence other members of the GPCR family). However, there is no evidence of homology between the linear sequences of the two families of proteins. It has been argued (Pardo *et al.*, 1992) that homology between bR and GPCRs can be detected if one allows for the membrane-spanning helices to be shuffled within the sequence (i.e. any helical region of a GPCR is allowed to align with any helical region of bR). The result is intriguing, based on the sequence of bR and aminergic GPCRs. However, if one extends the analysis to a comparison of all sequences clearly related to bR and all sequences from the GPCR family then the observed similarity breaks down. The lack of homology in the amino acid sequences, however, is not necessarily indicative of a lack of homology in the tertiary structures of these proteins. It is clear that two proteins with similar tertiary structures do not necessarily have detectable similarity in their amino acid sequences (Bajaj and Blundell, 1984). In addition to the lack of sequence homology there are also functional differences between bR and the eukaryotic rhodopsins. Bacteriorhodopsin is a proton pump and is not coupled

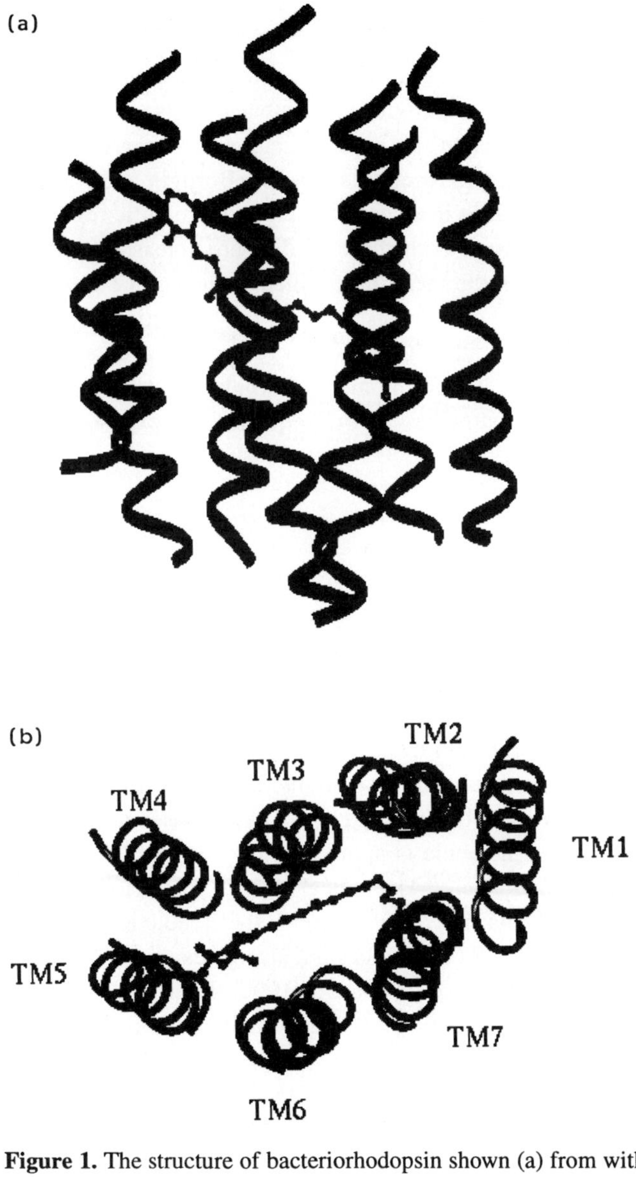

Figure 1. The structure of bacteriorhodopsin shown (a) from within the plane of the membrane and (b) viewed from above the extra-cellular surface of the membrane. The helices are labelled TM1–TM7 in the order in which they appear in the sequence of bacteriorhodopsin. TM, transmembrane.

to a G protein whilst rhodopsins are not known to be involved directly in the translocation of ions across membranes and are coupled to the G protein transducin.

Since there is no clear evidence of a relationship between bR and eukaryotic rhodopsins and therefore to other members of the GPCR family, it is dangerous to assume *a priori* that the tertiary folds are the same.

Membrane topography

The topography of the β2-adrenergic receptor and bovine rhodopsin have been defined using anti-peptide antibodies (Borjigin and Nathans, 1994, Wang *et al.*, 1989). Wang *et al.* raised antibodies to 11 peptide segments occurring in the hydrophilic regions of the β2 adrenergic receptor sequence. Borjigin and Nathans, conversely, inserted an epitope into sites in each of the eight hydrophilic regions of the rhodopsin sequence. The results of both these studies are consistent with an organization of these two proteins in the membrane which is shown in *Figure 2*. It is worth stressing that these data are also consistent with the prediction, based on hydrophobicity profiles (see below), that GPCRs have seven bilayer spanning elements and that these elements lie within the more hydrophobic stretches of the sequence, whereas the hydrophilic stretches of the sequence are exposed at the surface of the membrane.

Spectroscopic studies of rhodopsins

Infra-red linear dichroism studies of frog rhodopsin (Saibil *et al.*, 1978) indicate that this molecule, in the disc membrane of rod cells, has segments with an alpha-helical structure which are oriented approximately perpendicular to the plane of the membrane. Further infra-red and far ultraviolet circular dichroism studies of bovine rhodopsin (Rothschild *et al.*, 1980) also concluded that this molecule contained alpha helices and that these had a tilt with respect to the membrane normal of less than 40°. Recently a detailed characterization of bovine rhodopsin using Fourier transform infra-red spectroscopy (Garcia-Quintana *et al.*, 1994) has concluded that the content of alpha helix in this molecule is approximately 51% which includes regular helical structures and also 3_{10} helix. The helical regions appear to be largely inaccessible to solvent. These data not only support the prediction that the membrane spanning elements of GPCRs are helices but also gives us some clues as to the packing of the helices. Chothia and Finkelstein (1990) describe two general packing motifs for alpha helices, one in which the interaxial angle is approximately +20° and another in which the interaxial angle is approximately –50°. The observation from spectroscopic studies that the membrane spanning helices have a tilt no greater than 40° with respect to the membrane normal is most consistent with helix packing of the former class.

3. Evidence from theoretical analysis of GPCR sequences

Hydrophobicity profiles

Hydrophobicity profiles (Kyte and Doolittle, 1982) suggest the presence of seven membrane spanning segments in the sequences of GPCRs. The length of these segments is consistent with them being helices (i.e. they are of the order of 24 amino

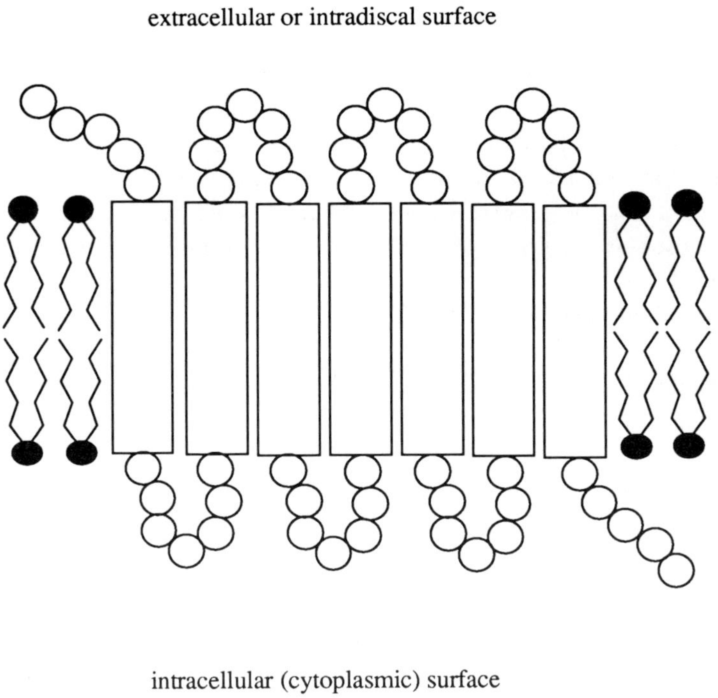

Figure 2. Expected topography of GPCRs in plasma membrane based on epitope mapping of the β_2-adrenergic receptor and bovine rhodopsin.

acids in length which is typical of the length of helices in bR). The profiles obtained in this way are consistent across an alignment of a divergent set of 24 sequences belonging to the GPCR family, as shown in *Figure 3*. Despite the sometimes low homology between GPCR sequences, the consistency in the hydrophobicity profiles is suggestive of gross similarity in their tertiary structures.

Conservation and hydrophobic moments

If a helical structure is assumed for stretches of GPCR sequences and the sequence plotted on helical wheels, it can be seen that moments for the hydrophobicity and conservation of residues appear on opposite sides of the wheel (Donnelly *et al.*, 1989). This suggests that each helix is exposed to two different environments (i.e. each of the seven membrane spanning helices in GPCRs lies on the boundary between two environments in which different constraints operate on residue variation patterns). One side of the helix is exposed to a hydrophobic environment assumed to be the lipid portion of the membrane and the other to an environment in

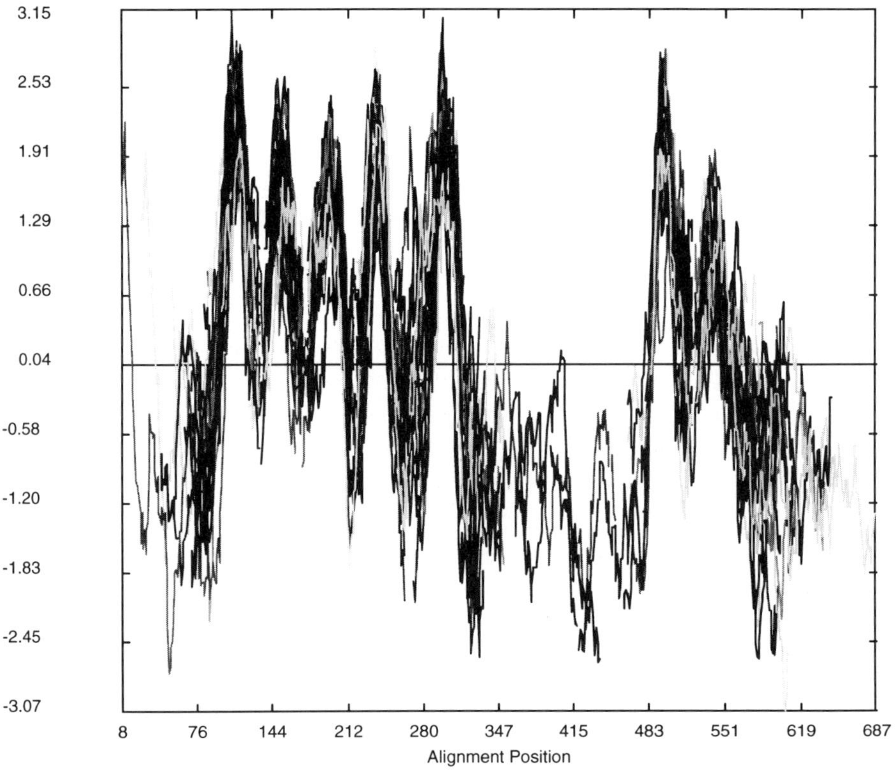

Figure 3. Kyte–Doolittle hydrophobicity profile of 24 aligned GPCR sequences generated using CAMELEON (published by Oxford Molecular Ltd). There is evidence for the presence of seven membrane spanning elements in all of the sequences. The vertical axis is hydrophobicity.

which conservation of key residues is important. In globular proteins, residues of structural and functional importance are those which are conserved over evolutionary time. If this observation is applied to GPCR sequences, it suggests that the conserved residues are of structural or functional importance. It would be logical to expect these residues to be associated with the core of the receptor where they could interact with ligands and one another. The occurrence of ligand binding within the core of the receptor is supported by the observation that deletion mutants of the β2-adrenergic receptor, in which the extramembraneous loops were removed, were still capable of binding ligands (Dixon *et al.*, 1987). Donnelly *et al.* (1994) have shown by analysis of multiple alignments of GPCR sequences that the periodicity in residue variance for those regions of the sequence which are thought to span the membrane is similar to the residue periodicity of a regular alpha helix (i.e. at approximately 100°).

4. Deriving a template for GPCR models

The experimental and theoretical evidence considered thus far suggest that the general structure of GPCRs is a seven helical membrane spanning bundle with the topography shown in *Figure 2* in which the helices are arranged in a ellipse. Each helix lies on the circumference of the ellipse exposed to two different environments. However, the position of each helix with respect to the others in the structure has not been considered.

In order to construct a three-dimensional model of a GPCR, a template representing the core structure of the receptor in which the spatial relationships of the membrane spanning helices with respect to one another is defined is required. A number of approaches to obtaining such a template can be considered; either the structure of bacteriorhodopsin can be used directly (e.g. Lewell, 1992) or indirectly (e.g. Findlay and Pappin, 1986; Livingstone *et al.*, 1992) or a template can be built *ab initio*, that is without using the co-ordinates of a known structure as a starting point (e.g. Maloney-Huss and Lybrand, 1992). The use of bR as a template either directly or indirectly implies the tertiary structures of GPCRs and bR are grossly the same, whilst the *ab initio* approach allows for different arrangements of the helices to be considered.

Validation of templates for modelling GPCRs

The template choosen should be consistent with the available experimental data. What follows is an attempt to derive the relative positions of the seven membrane spanning helices of GPCRs based on the available experimental data and theoretical arguments. Baldwin (1993) has used this approach combined with sequence analysis to derive a template for GPCRs.

In constructing a template for GPCR models, an important assumption will be made. This is that in gross terms the tertiary fold of all members of the GPCR family will be the same. The justification for this assumption is that in globular protein families it has been observed that a tertiary fold is preserved within families and superfamilies of proteins even amongst those family members with poor sequence homology to one another.

The two-dimensional projection map of bovine rhodopsin yields the approximate dimensions of the elliptical boundary within which the helical membrane spanning elements would be expected to fit. In using these dimensions the assumption is made that they are transferable to other GPCRs and that the shape of rhodopsin is dictated by the interaction between residues of the protein rather than the shape of the bound ligand, retinal. In defence of this argument it can be seen from sequence alignments that bovine rhodopsin shares a number of key residues with other members of the GPCR family (*Figure 4*). These conserved residues are, presumably, involved in either the structure or function of GPCRs and are therefore likely to maintain the same position with respect to one another in all GPCR structures. The projection map also indicates the positions at which helices should be placed within the envelope of the ellipse. Since the conserved residues of the GPCR family are distributed amongst the seven membrane spanning helices, the positions of the helices in other GPCRs can be expected to be in the same general location as seen in bovine rhodopsin The problem then reduces to a question of which helices should be placed at which positions within the elliptical envelope.

The first question which should be considered is whether the helices which are next to one another in the amino acid sequence are next to one another in the structure. This property of the helices will be called adjacency. If adjacency could be demonstrated for all seven helices the number of possible arrangements of the helices reduces from 5040 to 14.

The issue of adjacency has been addressed by considering the number of residues in the regions lying between the membrane spanning helices (Baldwin, 1993). The number of residues defined as lying in the interhelical region depends critically on the definition of which residues lie in the membrane spanning helices and which do not. These definitions can be inaccurate, resulting in uncertainty as to exactly which residue marks the start or end of each helix. Thus the number of residues assigned to the regions between the helical segments can easily be over- or under-estimated. However, using the helical regions defined from the hydrophobicity profiles and outlined in *Figure 4* and applying them to a multiple alignment of a large set of diverse GPCR sequences, figures for the maximum and minimum number of residues between helical segments can be estimated. These values are shown in *Table 1.* Also shown are the equivalent values for bacteriorhodopsin in which the beginnings and ends of helices can be approximated from the structure and in which all the helices are adjacent.

It has been suggested that where the number of interhelical residues is less than 10 that this indicates that the helices are adjacent. However, the question which needs to be asked is not 'What number of residues between sequential helices implies they are adjacent?' but 'What number of residues between sequential helices will allow two helices to be far enough apart to allow one or more helices to fit between them?'. This question cannot readily be answered from available structural data for membrane spanning helices. In bacteriorhodopsin the helices are all adjacent to one another. The only other suitable structure available for study is that of the photoreaction centre (Deisenhoffer and Michel, 1989). In this structure not all the membrane spanning helices are adjacent. The one non-adjacent pair, however, are separated by a loop of approximately 30 amino acids in length. Thus neither of the helical membrane spanning proteins of known structure are of use in determining what number of residues allow two membrane spanning helices to be non-adjacent.

```
Rhodopsin: MNGTEGPNFYVPFSNKTGVVRSPFEAPQYYLAEPWQFSMLAAYMFLLIML
                                                  hhhhhhhhhhh
Consensus:                                           !!  !! !!

Rhodopsin: GFPINFLTLYVTVQHKKLRTPLNYILLNLAVADLFMVFGGFTTTLYTSLH
           hhhhhhhhhhhhhh      hhhhhhhhhhhhhhhhhhhhhhhhh
Consensus:  ! gN!!!! !              !! la !d!! !!      !

Rhodopsin: GYFVFGPTGCNLEGFFATLGGEIALWSLVVLAIERYVVVCKPMSNFRFGE
                  hhhhhhhhhhhhhhhhhhhhhhhhhhh
Consensus:   w  g   C !          !%!  l !! !dRy !!  p

Rhodopsin: NHAIMGVAFTWVMALACAAPPLVGWSRYIPEGMQCSCGIDYYTPHEETNN
               hhhhhhhhhhhhhhhhhhhhhhh    *  * *    *****
Consensus:   ! ! !!!!W!! !!! !p !

Rhodopsin: ESFVIYMFVVHFIIPLIVIFFCYGQLVFTVKEAAAQQQESATTQKAEKEV
            hhhhhhhhhhhhhhhhhhhhhh                         hh
Consensus:         !! # !p !!    y  !   !                  $!

Rhodopsin: TRMVIIMVIAFLICWLPYAGVAFYIFTHQGSDFGPIFMTIPAFFAKTSAV
           hhhhhhhhhhhhhhhhhhhhhhhh       hhhhhhhhhhhhhhhhhhh
Consensus:    !!!!!! f !cw p                  !  !         %

Rhodopsin: YNPVIYIMMNKQFRNCMVTTLCCGKNPLGDDEASTTVSKTETSQVAPA
           hhhhhhh
Consensus: !nP!!y    $ f
```

Figure 4. The sequence of bovine rhodopsin. The consensus sequence derived from an alignment of diverse GPCR sequences is shown below the sequence (! = hydrophobic, # = aromatic, $ = basic, % = hydroxyl containing). The regions predicted to lie within the membrane spannig helices are marked by an 'h' and positions at which a cysteine occurs in the loop joining helices 4 and 5 are marked by an asterisk.

At present a crude analysis using globular protein structures has been undertaken in an attempt to gain some insight to this question.

The IDITIS database of protein structures (Thornton and Gardener, 1989) was searched for the occurrence of secondary structure motifs of the type end-of-helix-(non-helical residue)$_n$-start-of-helix, where n is the number of non-helical residues between the ends of two helices which follow one another in the amino acid sequence. The distance between the C_α of the last residue of the N-terminal helical section and the first C_α of the C-terminal helical section was recorded for values of n between 1 and 7. The minimum and maximum distances found between the ends of the helices are recorded in *Table 2*. The analysis does not take into account many of the factors which might influence the separation between two helical ends and was used only to assess how far apart the ends of two sequential helices could be, depending on the number of residues between their ends. Comparing the data in *Table 2* to measurements taken from the structure of bR it suggests that helices separated by one or two amino acids can be assumed to be adjacent to one another. Helices separated by three, four or five residues would be expected to be adjacent

Table 1. Minimum number of amino acid residues found between the ends of helices in the bacteriorhodopsin GPCR families

Interhelical regions	Minimum number of intervening residues in bacteriorhodopsin	Minimum number of intervening residues in GPCRs
1–2	5	5
2–3	11	11
3–4	5	17
4–5	7	3
5–6	7	16
6–7	10	5

or non-adjacent across the corner of the bounding ellipse. Finally, those helices separated by more than five residues could be adjacent, non-adjacent across the corner of the bounding ellipse, or non-adjacent along the long axis of the ellipse.

The data presented in *Table 1* suggests that it is not possible to assume adjacency for any of the GPCR helices, although it is likely that helices 4 and 5 (minimum separation three residues) are adjacent. Therefore the property of adjacency cannot be used to reduce significantly the number of possible arrangements of the helices.

The next constraint we might consider is that, except in the MSH receptor and its close homologues, there is a disulphide bond formed between the extracellular loop joining helices 4 and 5 and the N-terminus (extracellular end) of helix 3. The position of the cysteine in helix 3 is fully conserved whilst its partner cysteine appears at a number of sites within the loop joining helices 4 and 5 (marked with an asterisk in *Figure 4*). In some GPCRs this cysteine lies within five residues of the likely extracellular end of helix 4 or 5. This suggests that the N-terminal end of helix 3 is close to the extracellular ends of both helices 4 and 5 and supports the idea that helices 4 and 5 are adjacent to one another. This arrangement can best be accommodated on a corner of the ellipse and there are then four possible arrangements of the helices shown in *Figure 5*. The proximity of helix three to helix five is also supported by the observation that residues on both helices are important for the binding of agonists of the β2 adrenergic receptor (Strader *et al.*, 1988; 1989).

The next helix to consider is helix 6. The retinal analogue shown in *Figure 6* forms cross-links predominantly to residues on helices 3 and 6 of bovine rhodopsin (Nakayama and Khorana, 1990). These data suggest that the ionone head group of retinal is bound between helices 3 and 6 and that therefore helices 3 and 6 are likely to be opposite one another. If one further takes into account the recent data of Elling *et al.* (1995) showing that it is possible to create a zinc binding site in the human NK1 (neurokinin) receptor by introducing histidine residues into helices 5 and 6 then helix six is also likely to be adjacent or opposite to helix 5. In the arrangements

Table 2. Minimum and maximum separations of the terminal C_α atoms of helices with *n* residues separating their ends

Number of amino acids between helix ends	Minimum separation of helix ends (Å)	Maximum separation of helix ends (Å)
1	4.7	7.4
2	4.9	10.7
3	5.0	23.8
4	4.6	16.7
5	5.0	19.6
6	6.1	21.6
7	6.8	24.6

shown in *Figure 5*(*c*) and (*d*) it is not possible to place helix 6 adjacent to or opposite helix 5 and thus these two arrangements are considered to be unlikely candidates for a template and will not be considered further. There are four possible templates which can now be derived from the arrangements of helices shown in *Figure 5*(*a*) and (*b*); these are illustrated in *Figure 7*. Of these arrangements those illustrated in *Figures 7*(*b*) and (*c*) can be excluded since helix 6 is not opposite to helix 3.

Helix 7 may now be added to the pattern being built up. The experimental constraint used to position this helix relative to the others is data which suggest that the counter-ion to the Schiff's base formed between retinal and lysine 296 on helix 7 of bovine rhodopsin is glutamate 113 on helix 3 (Lin *et al.*, 1992). The direction of vectors associated with the side chains concerned can be estimated by constraining the hydrophobic moments of helices 3 and 7 to point to the outside of the receptor and the helix axis to lie approximately parallel to the membrane normal. Templates consistent with these data should have the vectors of both side chains pointing towards the same point. Adding helix 7 to the two remaining arrangements of helices in *Figure 7* yields the arrangements shown in *Figure 8*. It can be seen from *Figure 8*(*a*) that adding helix 7, in the appropriate location with respect to helix 3, to the template in *Figure 7*(*d*) places helix 7 outside the boundary of the receptor. This leaves the possible templates illustrated in *Figures 8*(*b*) and (c). The arrangement shown in *Figure 8*(*c*) is unlikely since they place helices 6 and 7 in a position which would be difficult to link with five residues (see *Table 1*). The arrangement shown in *Figure 8*(*b*), therefore, represents the most likely arrangement of helices 3 to 7.

There is currently no experimental evidence which would allow the two possible arrangements of the remaining two helices, one and two, to be distinguished from one another (see *Figure 9*). However, based on the calculated relative hydrophobic surface area of these two helices (Baldwin, 1993) the arrangement shown in *Figure 9*(*a*) is the most probable. This arrangement of helices is the same as observed in bacteriorhodopsin and agrees with the conclusions reached by Baldwin. It also turns

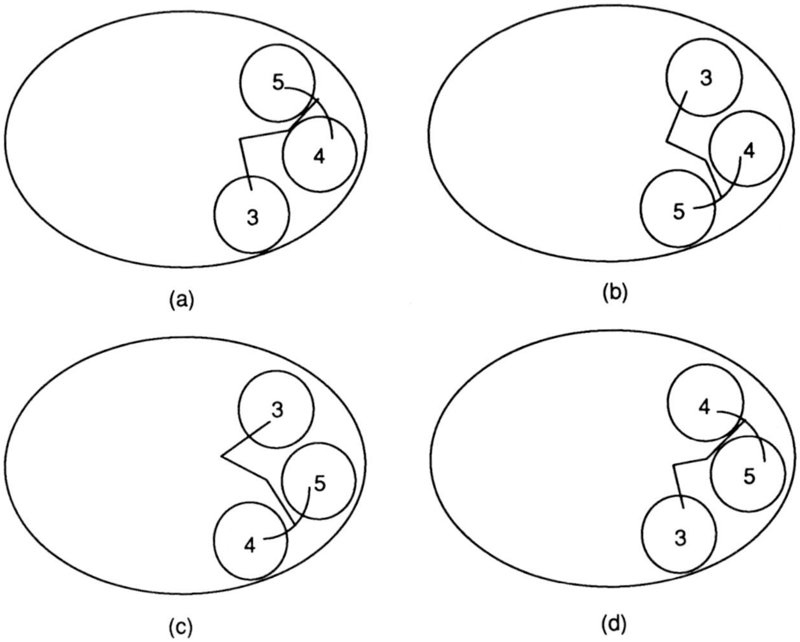

Figure 5. The four possible arrangements of helices 3, 4 and 5 in a GPCR model template. The helices are represented as circles, the loop joining helices 4 and 5 as an arc and the disulphide bridge as a zig-zag.

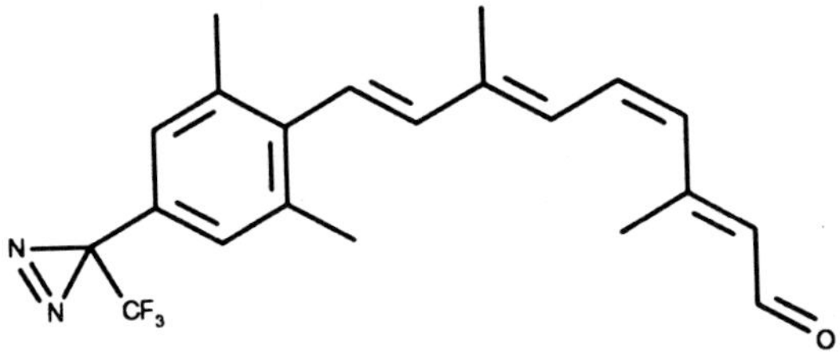

Figure 6. The structure of *o*-dimethyl *p*-trifluoromethyldiazirine phenyl retinal, a retinal analogue used for cross-linking studies in bovine rhodopsin.

out that this arrangement of helices allows those residues which are well conserved amongst members of the GPCR family to form clusters, which is what would be expected if they are of structural and functional significance. Thus, overall, it would appear that this template is in agreement with most of the available experimental information.

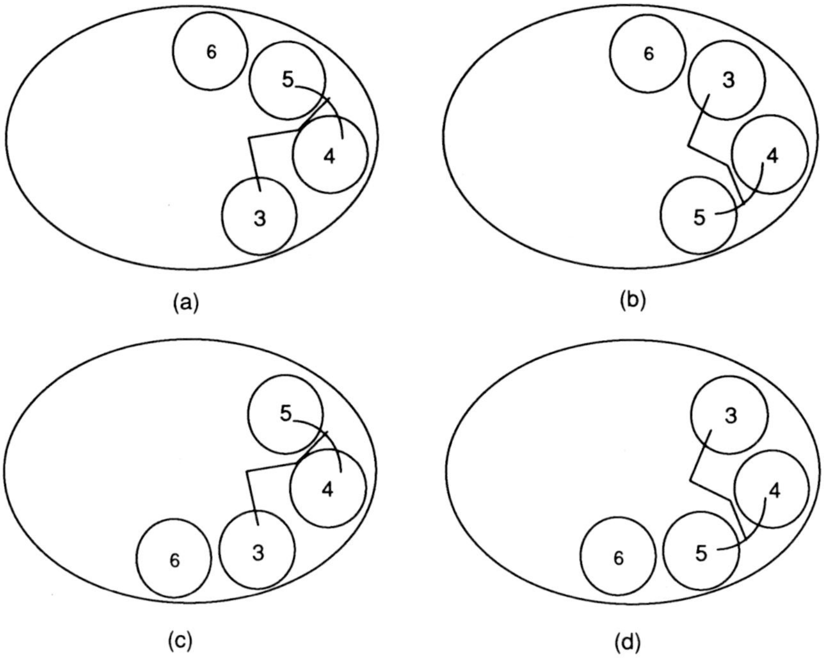

Figure 7. The four possible arrangements of helices 3, 4, 5 and 6 in a GPCR model template. The helices are represented as circles.

Having established that this arrangement of helices is appropriate for models of G protein-coupled receptors, it is then possible either to build a model using the co-ordinates of bacteriorhodopsin as a template or to build a model *ab initio*. There are difficulties with both approaches.

Using bacteriorhodopsin as a template means that structural features which are peculiar to this protein will be reproduced in the model. Since the structure of only one seven helical membrane spanning protein has been determined it is difficult to know what features are likely to be unique to that structure. The projection map of bovine rhodopsin and the comparison with projections maps of bacteriorhodopsin at the equivalent resolution (Schertler *et al.*, 1993) do suggest that there are differences between the two structures. So, whilst the overall tertiary folds are expected to be the same based on this analysis, it is unlikely that the structures will be the same at a more detailed level. Structural variation is also likely between members of the GPCR family itself.

Building a model *ab initio* necessitates a model being constructed for each of the membrane spanning helices and then these helices being arranged to give a tertiary fold. This approach allows the possible variations between GPCRs and bR and

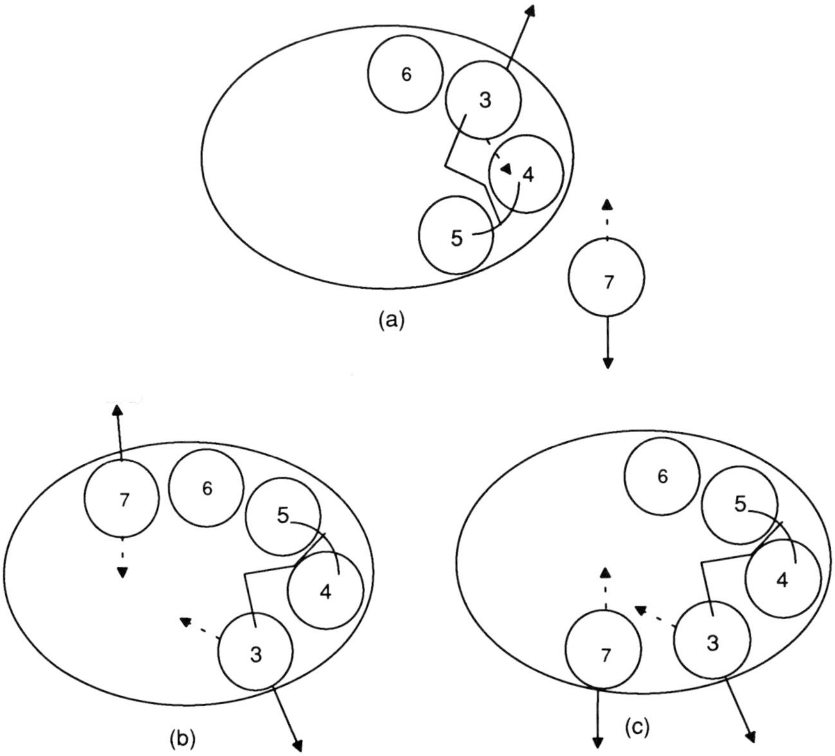

Figure 8. The four possible arrangements of helices 3, 4, 5, 6 and 7 in a GPCR model template. The side chain vectors of lysine 396 (helix 7) and glutamate 113 (helix 3) are indicated with arrows (dashed lines) as are the approximate hydrophobic moments of helices 3 and 7 (solid lines).

between different GPCRs to be taken into account. However, other problems need to be considered. These include the means by which the individual helices are constructed taking into account any global or local deviations from ideal helical structure and how the helices will subsequently be docked together.

Matching the sequence to the template

Once the template has been obtained the sequence of the GPCR being modelled must be matched to the residues of the template. The stretches of sequence involved can be identified from a hydrophobicity profile. However, the profile does not give the precise residues at which any membrane spanning helix begins and ends so some adjustment may often be necessary. In addition, the template used (whether bR or an *ab initio* template) cannot be aligned by residue identities to the sequence of the GPCR to be modelled. A possible solution is to match the sequence to be modelled to the template by considering the properties of the residues to be placed on the

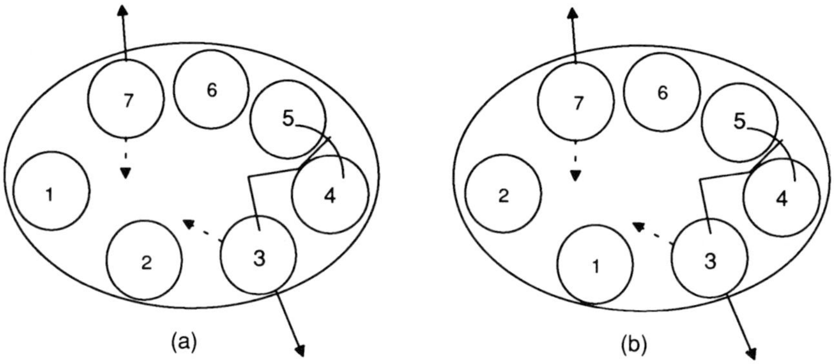

Figure 9. The two possible arrangements of helices 1–7 in a GPCR model template. The helices are represented as circles.

template in relation to their possible positions in the template. One way of performing this match is to plot the sequence in question on a helical wheel. This will reveal the likely lipid facing and pore facing residues of the helix using this information the sequence can then be transferred to the template. However, the match is rarely perfect and adjustment of the sequence matches and the local structure of each helix may need to be considered, for example the presence of prolines within a helix may require adjustment of the helix structure.

Modelling the loops which link the membrane spanning helices

Many GPCR modellers avoid the problem of building loops by simply omitting them from their models. Since it is difficult to model the loops correctly from globular proteins which are greater than about 5–6 residues in length, it is unlikely that GPCR loops can be modelled correctly especially in the light of the anticipated error in correct positions of the membrane spanning helices which define the anchor points to which a loop might be fitted. It is, therefore, highly likely that the conformation of modelled loops in GPCR models will be incorrect and of little use in interpreting such models.

5. Conclusions

It appears that a satisfactory template describing the tertiary fold of GPCRs can be derived using the available experimental information and some theoretical arguments. However, it is also clear that whether models are constructed *ab initio* or on the basis of the structure of bacteriorhodopsin that there is potential for error in the model. The expected error makes the specification of exact interactions between residues in the receptor or between the receptor and a ligand difficult since the model cannot hope to represent the true positions of all atoms in the receptor. At best it can suggest which interactions are possible and therefore worthy of experimental investigation. In fact, the best way to use such models is as a means of generating hypotheses which can be directly tested by experiment. It is only by constantly testing our models against empirical data that they and the methods by which they are generated and interpreted can be improved.

References

Baldwin JM. (1993) The probable arrangement of the helices in G-Protein coupled receptors. *EMBO J.* **12**: 1693–1703.

Bajaj M, Blundell TL. (1984) Evolution and the tertiary structure of proteins. *Ann. Rev. Biophys. Bioeng.* **13**: 453–492.

Borjigin J, Nathans J. (1994) Insertional mutagenesis as a probe of rhodopsins topography, stability and activity. *J. Biol. Chem.* **269**: 14715–14722.

Chothia C, Finkelstein A. (1990) The classification and organisation of protein folding patterns. *Ann. Rev. Biochem.* **59**: 1007–1039.

Corless JM, McCaslin DR, Scott BL. (1982) Two-dimensional rhodopsin crystals from disk membranes of frog retinal rod outer segments. *Proc. Natl Acad. Sci.* **79**: 1116–1120.

Cronet P, Sander C, Vriend G. (1993) Modelling of transmembrane seven helix bundles. *Prot. Eng.* **6**: 59–64.

Deisenhofer J, Michel H. (1989) The photosynthetic reaction centre from the purple bacterium *Rhodopseudomonas viridis*. *EMBO J.* **8**: 2149–2170.

Dixon RAF, Sigal IS, Rands E, Register RB, Candelore MR, Blake AD, Strader CD. (1987) Ligand binding to the beta-adrenergic receptor involves its rhodopsin-like core. *Nature*, **326**: 73–77.

Donnelly D, Johnson MS, Blundell TL, Saunders J. (1989) An analysis of the periodicity of conserved residues in the sequence alignments of G-protein coupled receptors. *FEBS Lett.* **251**: 109–116.

Donnelly D, Overington JP, Blundell TL. (1994) The prediction and orientation of alpha helices from sequence alignments: The combined use of environment-dependent substitution tables, Fourier transform methods and helix capping rules. *Prot. Eng.* **7**: 645–653.

Elling CE, Nielsen SM, Schwartz TW. (1995) Conversion of antagonist-binding site to metal-ion site in the tachykinin NK-1 Receptor. *Nature*, **374**: 74–77.

Findlay JBC, Pappin DJC. (1986) The opsin family of proteins. *Biochem. J.* **238**: 625–642.

Garcia-Quintana D, Garriga P, Manyosa J. (1994) Quantitative characterization of the structure of rhodopsin in disc membrane by means of Fourier transform infra-red spectroscopy. *J. Biol. Chem.* **268**: 2403–2409.

Henderson R, Baldwin JM, Ceska TA, Zemlin F, Beckmann E, Downing KH. (1990) Model for the structure of bacteriorhodopsin based on high-resolution electron cryomicroscopy. *J. Mol. Biol.* **213**: 899–929.

Hibert MF, Trumpp-Kallymeyer S, Bruinvels A, Hoflack J. (1991) Three-dimensional models of neurotransmitter G-binding protein coupled receptors. *Mol. Pharmacol.* **40**: 8–15.

Kyte J, Doolittle RF. (1982) A simple method for displaying the hydropathic character of a protein. *J. Mol. Biol.* **157**: 105–132.

Lewell XQ (1992) A model of the adrenergic beta-2 receptor and binding sites for agonist and antagonist. *Drug Design Discovery*, **9**: 29–48.

Lin SW, Sakmar TP, Franke RR, Khorana HG, Mathies RA. (1992) Resonance Raman microprobe spectroscopy of rhodopsin mutants: effects of substitutions in the third transmembrane helix. *Biochemistry*, **31**: 5105–5111.

Livingstone CD, Strange PG, Naylor LH. (1992) Molecular modelling of D2-like dopamine receptors. *Biochem. J.* **287**: 277–282.

Maloney-Huss K, Lybrand TP. (1992) Three-dimensional structure for the β_2 adrenergic receptor protein based on computer modelling studies. *J. Mol. Biol.* **225**: 859-871.

Nakayama TA, Khorana HG. (1990) Orientation of retinal in bovine rhodopsin determined by cross-linking using a photoactivatable analog of 11-cis retinal. *J. Biol. Chem.* **265**: 15762–15769.

Pardo L, Ballesteros JA, Osman R, Weinstein H. (1992) On the use of the transmembrane domain of bacteriorhodopsin as a template for modelling the three-dimensional structure of guanine nucleotide-binding regulatory protein-coupled receptors. *Proc. Natl. Acad. Sci. USA*, **89**: 4009–4012.

Rothschild KJ, Sanches R, Hsiao TL. (1980) A spectroscopic study of rhodopsin alpha-helix orientation. *Biophys. J.* **31**: 53–64.

Saibil H, Michel-Villaz M, Chabre M. (1978) Orientation of alpha-helical segments in frog rhodopsin studied by infra-red linear dichroism. *Biophys. J.* **21**: 178.

Schertler GF, Villa C, Henderson R. (1993) Projection structure of rhodopsin. *Nature*, **362**: 770–772.

Strader CD, Sigal IS, Candelore MR, Rands E, Hill WS, Dixon RAF. (1988) Conserved aspartic acid residues 79 and 113 of the β-adrenergic receptor have different roles in receptor function. *J. Biol. Chem.* **263**: 10267–10271.

Strader CD, Candelore MR, Hill WS, Sigal IS, Dixon RAF. (1989) Identification of two serine residues involved in agonist activation of the beta-adrenergic receptor. *J. Biol. Chem.* **264**: 13572–13578.

Thornton JM, Gardener SP. (1989) Protein motifs and database searching. *Trends Biochem. Sci.* **14**: 300–304.

Wang H, Lipfert L, Malbon CC, Bahouth S. (1989) Site-directed anti-peptide antibodies define the topography of the β-adrenergic receptor. *J. Biol. Chem.* **264**: 14424–14431.

Chapter 6

Pride and prejudice and G protein-coupled receptor models

Friedrich Rippmann

1. G protein-coupled receptors

G protein-coupled receptors (GPCRs) play a central role in the transmission of exogenous signals through the cell membrane. Their exogenous ligands range from small molecules such as dopamine or serotonin to peptides such as angiotensin II and proteins such as interleukin-8. They are involved in many diverse processes, such as colour vision, odour perception, blood pressure regulation, sexual activity, depression, schizophrenia, and neutrophil activation. Many members of this large family are of prime pharmacological interest.

The results of various experimental techniques which have established a general topography for GPCRs have been reviewed recently (Savarese and Fraser, 1992). The N- and C-termini are extra- and intracellular, respectively. GPCRs vary greatly in length. Seven hydrophobic stretches span the membrane connected by loops ranging in length from 5 to 420 residues (Baldwin, 1993). The second and third intracellular loops and the cytoplasmic C-terminus are involved in coupling to the G-protein (Fraser *et al.*, 1988; Lechleiter *et al.*, 1990; Strader *et al.*, 1989). Mutagenesis experiments suggest that the region of the membrane spanning helices is where small ligands bind. This corresponds to the position of the retinal moiety in several photosensitive GPCRs. The number of different GPCRs in any one species is likely to be very large. Often, compounds active on one of these receptors have a mixed profile of activities which reflects their interaction with a multitude of structurally similar but distinct receptors.

2. Structures of integral membrane proteins

The X-ray-derived structures of porin PhoE, the outer membrane protein from *E. coli*, and of the photosynthetic reaction centre from the purple bacterium *Rhodopseudomonas viridis* (Deisenhofer and Michel, 1989), and the electron diffraction structure of bacteriorhodopsin (Henderson *et al.*,1990) were used to derive some general features for the interaction of integral membrane proteins with the phospholipid bilayer (lipopolysaccharide in the case of the porins).

The porin structure in particular, and to a lesser extent the photosynthetic reaction centre structure, show a clear preference for Tyr and Trp residues at the inner interface between the hydrophobic fatty acids and polar headgroups of the membrane lipids (*Figure 1,* see page 105). Further away from the centre of the lipid bilayer there is a stripe of Arg and Lys residues which might interact at the aqueous side of the membrane with (charged) headgroups of the phospholipids.

Following analysis of bacterial membrane proteins, von Heijne derived the 'positive inside rule' for membrane spanning proteins (Heijne, 1986). It is interesting to note that this feature is obviously independent of the overall fold of the protein: porins are all-beta structures, whereas the membrane spanning parts of the photosynthetic reaction centre and bacteriorhodopsin are all-helical. The 'positive inside rule' also seems to be applicable at least to some transmembrane segments in human GPCRs (Rippmann, 1994). These features were helpful in the identification of the ends of the helices.

An additional common feature of the photosynthetic reaction centre and of bacteriorhodopsin to bear in mind is a 20–25° tilt of the membrane spanning helices.

3. Sequence analysis of GPCRs

All available sequence information on GPCRs was collected from publicly available sequence databases (i.e. PIR (Barker *et al.*, 1991) and SWISSPROT (Bairoch and Boeckmann, 1991)). As a first step, the sequence of the 5-HT_2-receptor (5-hydroxytryptamine, 5-HT, serotonin) (Saltzman *et al.*, 1991) was used to collect related sequences of GPCRs from the SWISSPROT database using FASTA (Pearson,1990). The PILEUP program implemented in the GCG package (Genetics Computer Group, Inc., Madison, Wisconsin, USA) was used for the generation of a first preliminary multiple sequence alignment. The alignment parameters 'gap penalty' and 'gap length penalty' were modified to improve the alignment (the 'gap length penalty' was finally set to zero to ensure the alignment of GPCRs with large and small extracellular domains). Using the method of Gribskov *et al.* (1987) a common sequence profile was derived. This was then used to scan the SWISSPROT and PIR databases for more related sequences of GPCRs using Genetics Computer Group's PROFILESEARCH. This procedure should ensure a complete though redundant collection of relevant sequences. Redundancies between the sequences found in the two databases, and sequences without a pharmacological relevance, were removed manually. The programs MALIGNED (Clark, 1991) and GDE (Smith, 1990) were used for the visualization of the conserved regions in the alignments in particular. The result of such an alignment of all human (and a few non-human) GPCR sequences is shown in part in *Figure* 2.

It can be seen in *Figure 2* that the automatic procedures do not always lead to a satisfactory alignment. A cysteine at the beginning of helix 3 which is involved in a conserved disulphide bridge is not aligned correctly in all sequences. In order to keep the results presented here reproduceable, however, no manual readjustment of the multiple sequence alignment was carried out.

Another interesting feature of the loop connecting helices 4 and 5 (which usually contains one of the Cys residues of the conserved disulphide bond which bridges to the Cys at the start of helix 3) is its extreme shortness in the few receptors which do not contain this disulphide bridge (e.g. the melanocortin receptors at the top of multiple sequence alignments). This suggests that a certain loop length is necessary to bring the Cys residues into spatial proximity, and provides another useful feature for the derivation of the three-dimensional models.

Various approaches were used to identify the membrane spanning segments. Close inspection of the conserved and variable regions can reveal the position of the seven transmembrane helices from the multiple sequence alignment alone: *Figure 2* shows the seven well-conserved, deletion/insertion-free blocks which are separated by variable deletion/insertion-rich regions. Similar results can be derived from the analysis of hydropathy plots using an appropriate window of 20 amino acids for the averaging (Engelman *et al.*, 1986).

The model recognition approach to the prediction of all-helical membrane structure and topology of Jones *et al.* (1994) (program MEMSAT) was used to predict the membrane spanning segments in all individual sequences of human GPCRs, and the method of Persson and Argos (1994) (program TMAP) was applied to the multiple sequence alignment used here. In most cases, the predicted segments correspond very well with the conserved portions of the multiple sequence alignment (*Figure 2*), however, the TMAP prediction fails to identify the last transmembrane helix. These approaches narrow the likely ends of the helices to a range of just a few residues. *Figure 2* shows part of the multiple sequence alignment and the predictions (the dark 'xxxx' bars under the respective sequence are the MEMSAT predictions; the dark/light alternating 'xzxz' bar above the first sequence in the multiple sequence alignment is the TMAP prediction).

4. Three-dimensional models of GPCRs

Recently, numerous three-dimensional models of the transmembrane regions of various GPCRs have been published, of which only a selection can be mentioned here (Cronet *et al.*, 1993; Donnelly *et al.*, 1994; Hibert *et al.*, 1991; MaloneyHuss and Lybrand, 1992; Oliveira *et al.*, 1993a; Oliveira, *et al.*, 1993b; Rippmann and Boettcher, 1993; Roeper *et al.*, 1994; Trumpp-Kallmeyer *et al.*, 1992; Westkaemper and Glennon, 1993). However, in most cases the coordinates are not available, and the published details of the model building procedure are usually not sufficient to reproduce the models.

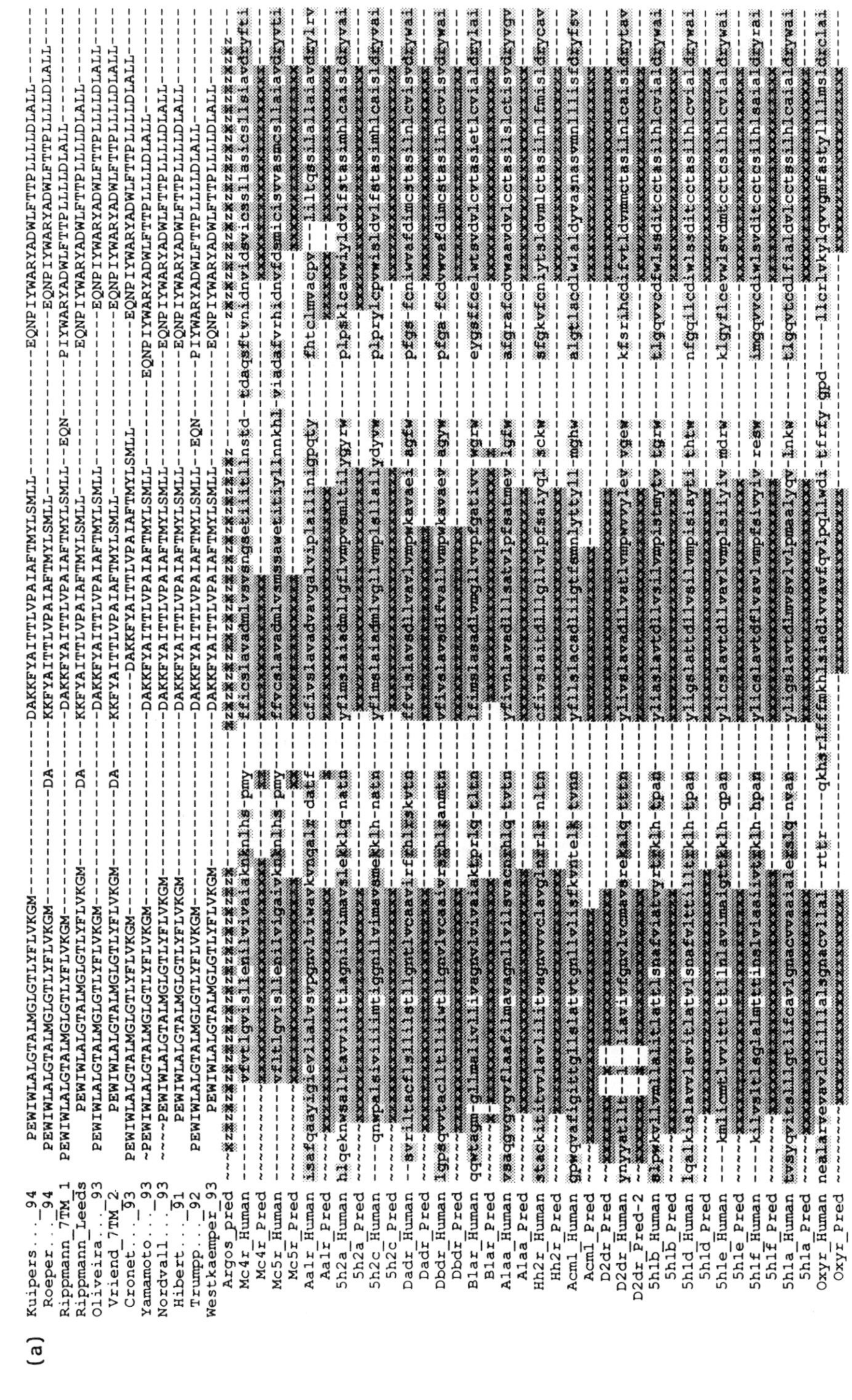

Figure 2(a). See page 97 for legend.

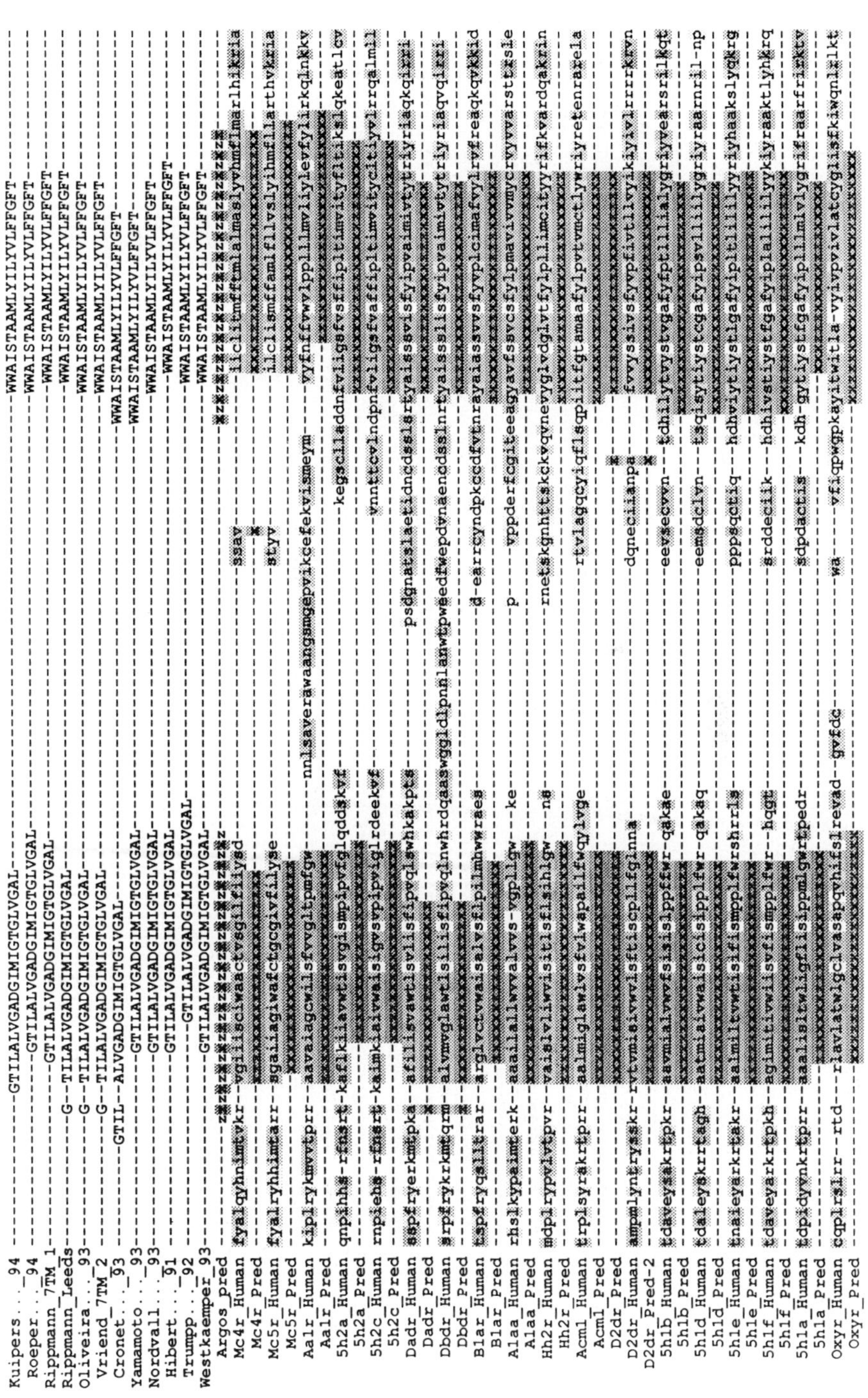

Figure 2(b). See page 97 for legend.

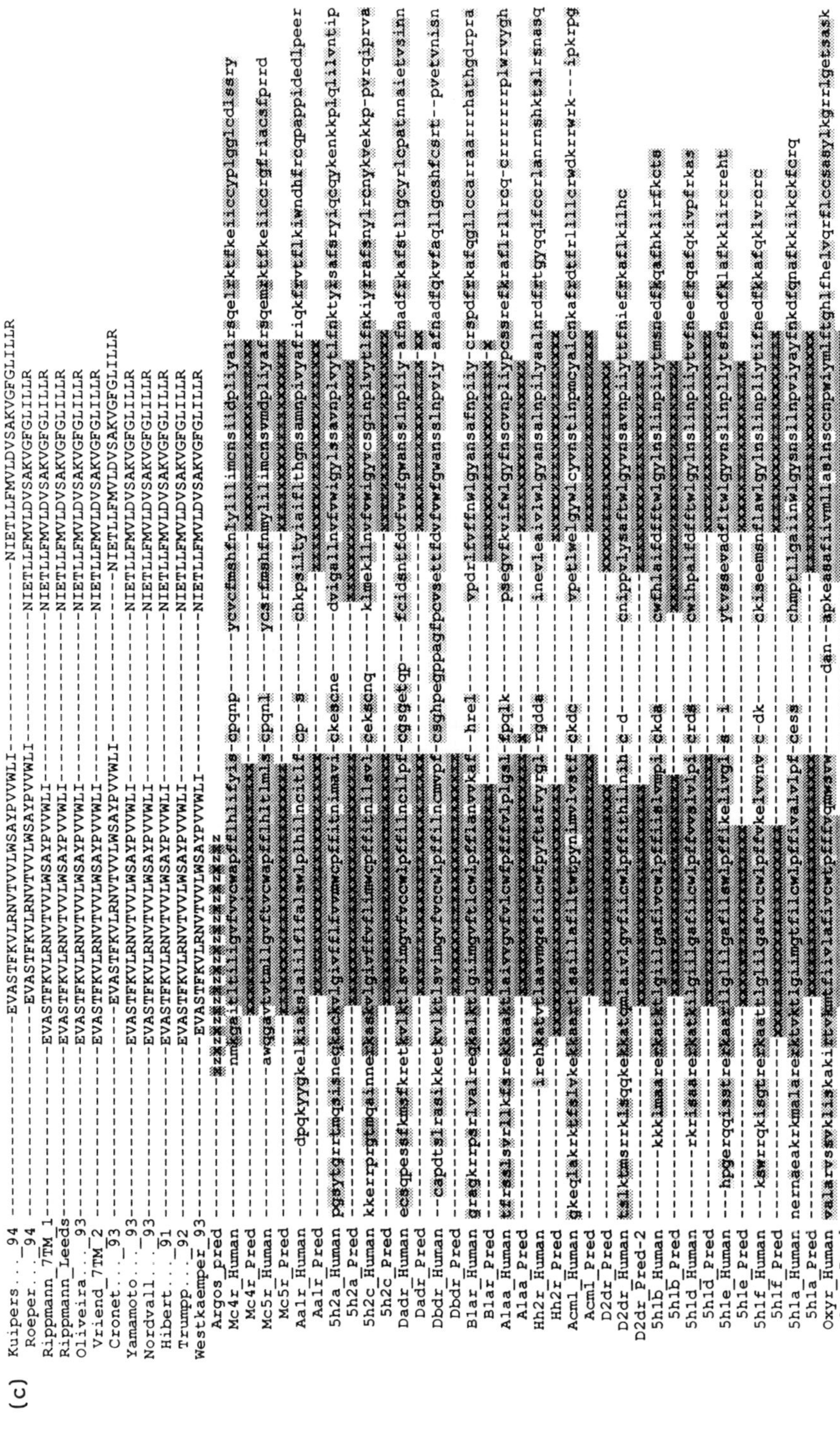

Figure 2(c). See page 97 for legend.

Figure 2. Part of a multiple sequence alignment of all human GPCRs. (a) Helices 1–3, (b) helices 4, 5, (c) helices 6, 7 (the N- and C-terminal regions and the large insertion between helices 5 and 6 are not shown). The top 12 lines are not part of the automatic alignment. They show the alignment of different GPCR sequences to bacteriorhodopsin as derived by various authors (Kuipers *et al.*, 1994; Roeper *et al.*, 1994; Rippmann and Boettcher, 1993; Rippmann–Leeds, this article; Oliveira *et al.*, 1993a; Oliveira *et al.*, 1993b; Cronet *et al.*, 1993; Yamamoto *et al.*, 1993; Nordvall and Hacksell, 1993; Hibert *et al.*, 1991; Trumpp-Kallmeyer *et al.*, 1992; Westkaemper and Glennon, 1993). The dark 'xxxx' bars indicate the MEMSAT (Jones *et al.*, 1994) predictions for each GPCR sequence, the dark/light alternating 'xzxz' bar indicates the TMAP (Persson and Argos, 1994) prediction, which is based on multiple sequences. The grey shading underlying the sequences was to highlight various properties of the multiple sequence alignment.

The model presented and discussed here was generated in an iterative manner using (a) the experimental bacteriorhodopsin structure as template, (b) the predicted helix positions in the GPCR sequences as derived above, (c) mutagenesis data and (d) docking of various preferentially rigid ligands (*Figure 3*) into the putative binding site. It is known from mutagenesis experiments and ligand attachment in rhodopsin that at least some of the ligands bind in the region of the seven transmembrane helices of the GPCRs. The only experimentally determined structure of a membrane protein with seven transmembrane helices is bacteriorhodopsin (Henderson *et al.*, 1990). The intra- and extracellular domains are not resolved in the experimental structure.

Our understanding of the rules governing the packing of helices in a membrane is still rather limited (Swindells and Thornton, 1991) and therefore a template-free *ab initio* approach was not used here. However, it should be mentioned that such *ab initio* derived models have been published recently (Taylor *et al.*, 1994). In this situation it was felt safer to base the derivation of the three-dimensional models on the experimental electron diffraction structure of bacteriorhodopsin. This protein is a bacterial proton pump, not a G protein-coupled receptor. However, there are functional and structural relationships between bacteriorhodopsin and mammalian opsins which give some justification for the use of this structure (see e.g. Trumpp-Kallmeyer *et al.*, 1992). This stucture unfortunately is of rather low resolution, about 3.5 Å in x and y directions (plane of the membrane), about 10 Å in z (across the membrane). The coordinates from the Protein Data Bank (Bernstein *et al.*, 1977), entry 1BRD (Henderson *et al.*, 1990) were modified by shifting helix 4 by +3 Å (i.e. ~60% of the pitch of an α-helix) in the z-direction (R. Henderson, personal communication). The helical regions of bacteriorhodopsin were aligned with an annotated multiple sequence alignment of GPCRs (dark 'xxxx' bars in *Figure 2* indicate the MEMSAT predictions) with the aid of the multiple sequence alignment programs PILEUP and MULTAL (Taylor, 1987) and manual adjustment.

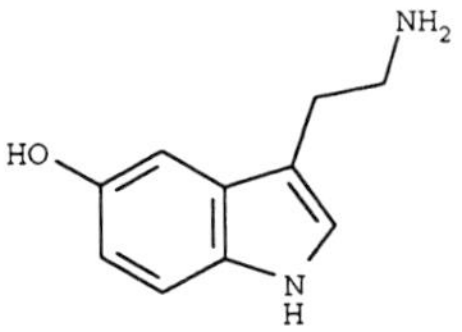

5-HT / Serotonin

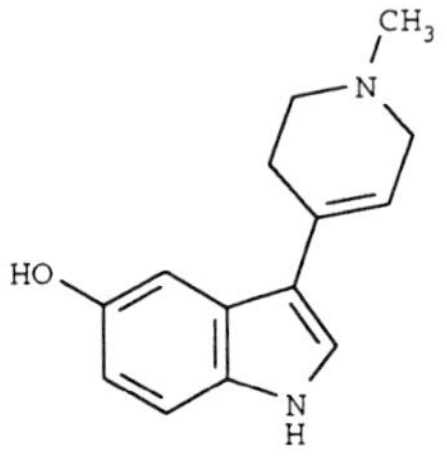

Methyltetrahydropyridinylindole

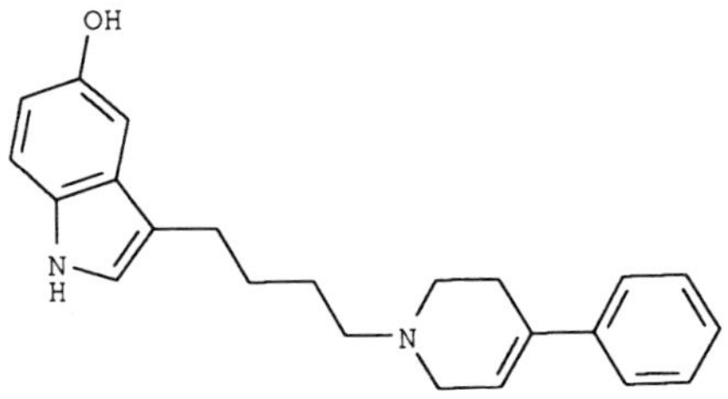

Roxindole

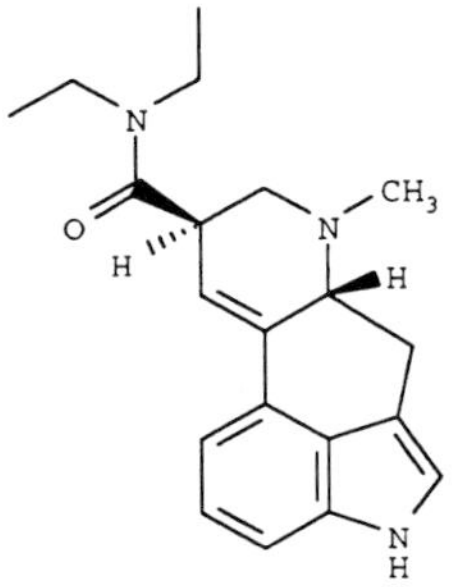

LSD
D-Lysergic acid diethylamide

Figure 3. Structural formulae of selected serotonergic agonists: 5-HT, methyltetrahydropyridinylindole, roxindole, *d*-LSD.

It is interesting to note, that, despite of the lack of overall sequence homology between bacteriorhodopsin on the one hand and GPCRs on the other, a suggestive sequence alignment can be generated for helices 6 and 7 between bacteriorhodopsin, mammalian opsins, and various other GPCRs (Rippmann, 1994). This may have led to the almost perfect unanimity in the alignment of the GPCR sequences to bacteriorhodopsin helices 6 and 7 by various authors (Cronet *et al.*, 1993; Hibert *et al.*, 1991; Kuipers *et al.*, 1994; Nordvall and Hacksell, 1993; Oliveira *et al.*, 1993a; Oliveira *et al.*, 1993b; Rippmann and Boettcher, 1993; Rippmann–Leeds, this article; Roeper *et al.*, 1994; Trumpp-Kallmeyer *et al.*, 1992; Westkaemper and Glennon, 1993; Yamamoto *et al.*, 1993); (top 12 lines in *Figure 2*), which is in sharp contrast to the alignments for all other helices where there is much less agreement between the various authors. In the case of helix 7, it is tempting to assume that the Lys residue which carries the retinal both in bacteriorhodopsin and mammalian opsins occupies the same position in the helix. However, the prediction by Cronet *et al.* (1993) which is rather independent of individual residue correspondences, differs from the other authors, and is in better agreement with the MEMSAT predictions (*Figure 2*).

Another interesting, though not really convincing correspondence is between the interhelical loops of bacteriorhodopsin and those of the GPCRs where in both groups the longest intracellular loop is the one connecting helices 5 and 6.

Because of their special pharmacological relevance in several central nervous system diseases, various 5-HT receptors were studied in detail. A model for the receptor 5-HT_{1A} was constructed using the homology building tools in WHATIF (Vriend, 1990). Visual inspection of the model and inclusion of a variety of mutagenesis data from the literature, (reviewed e.g. in Trumpp-Kallmeyer *et al.*, 1992 and Baldwin, 1993) and our analysis of ligand specificities (especially rigid ligands like tetrahydropyridinylindole or *d*-LSD (lysergic acid diethylamide), see *Figure 3*, were used to modify the initial alignment and three-dimensional model. Rough energy minimization was performed only for regions of interest (all backbone atoms fixed) using the SYBYL force field (Tripos Associates, Inc., St. Louis, Missouri, USA).

The X-ray structures of 5-HT (Amit *et al.*, 1978) and *d*-LSD (Baker *et al.*, 1972) were taken from the Cambridge Crystallographic Data Files (Allen *et al.*, 1983). The tetrahydropyridinylindole derivative was built in SYBYL. In this compound, the serotonin side chain flexibility is restricted to a narrow rotational cone. For structural formulae see *Figure 3*. The ligands were docked manually into the model using predetermined favourable positions for the phenolic hydroxy group and the charged side chain nitrogen. Asp-116 and Thr-199 were used to guide the docking. The importance of these residues for ligand binding has been shown for the β-adrenergic receptor by site directed mutagenesis (Strader *et al.*, 1989). The functionalities for the analysis of side chain interactions implemented in WHATIF were used to find suitable positions for an Asp/charged nitrogen interaction (paradigm: Asp in helix, contacting Lys in any context, see *Figure 4*, page 105) and for a phenolic OH/Thr-OH interaction (Tyr was used as a 'phenolic OH' substitute for the serotonin moiety).

E. Merck's roxindole (Boettcher *et al.*, 1992) (see structural formula in *Figure 3*), a potent 5-HT_{1A} agonist, dopamine agonist, and 5-HT reuptake inhibitor (Seyfried *et al.*, 1989) was also evaluated in the context of the model. This compound shows great promise in the treatment of major depression (Benkert *et al.*, 1992).

Roxindole is considerably larger than, for example, serotonin. Its flexibility is constrained to a chain segment connecting two rather rigid parts of this molecule. A U-shaped conformation was derived from a pharmacophore model resulting from the superposition of various 5-HT_{1A} agonists, for example serotonin, 8-OH-DPAT and *d*-LSD (results not shown). This conformation is similar to the results of a modelling study undertaken to explain the dopaminergic D2 activity of roxindole (Boettcher *et al.*, 1992). This conformation brings the tertiary nitrogen in close contact with the essential aspartate and leads to good hydrophobic contacts for the

4-phenyl-1,2,3,6-tetrahydropyridine ring system. *Figure 5* (see page 106) shows roxindole in this conformation in the context of the 5-HT_{1A} receptor.

Special attention was given to the different activities of *d*- and *l*-LSD. *d*-LSD binds to the 5-HT_{1A} receptor in the low nanomolar range, whereas *l*-LSD is still inactive at a 10 000-fold higher concentration. (A large compilation of affinity values can be found in Zifa and Fillion, 1992). This difference in activity corresponds to a possible hydrogen bond between the peptidic bond in *d*-LSD and a Ser residue on helix 4 (*Figure 6*, see page 106). This bond could not be formed by the inactive *l*-LSD stereoisomer.

5. Conclusions

Pride

Many models of various GPCRs have been published. Some of them use the experimental bacteriorhodopsin structure as a starting point (Cronet *et al.*, 1993; Hibert *et al.*, 1991; Kuipers *et al.*, 1994; Nordvall and Hacksell, 1993; Oliveira *et al.*, 1993a; Oliveira *et al.*, 1993b; Rippmann and Boettcher, 1993; Rippmann-Leeds, this article; Roeper *et al.*, 1994; Trumpp-Kallmeyer *et al.*, 1992; Westkaemper and Glennon, 1993; Yamamoto *et al.*, 1993); others do not (Donnelly *et al.*, 1994; MaloneyHuss and Lybrand, 1992; Taylor *et al.*, 1994). For the bacteriorhodopsin-based models, the overview of the underlying alignments (*Figure* 2) shows that all these models must differ in more than just fine detail. Nevertheless, most authors claim that they can explain various experimental observations with their models satisfactorily, and sometimes base far-reaching hypotheses on their models. This may sound more critical and surprising than it in fact is, because some models may be more similar than the alignments suggest, as a uniform shift of the helical sequences into one topographic direction ('in' or 'out') might lead to similar relative interactions between the helices. In fact, it can be seen from the alignments that most authors agree on the position of the helical stretches in the sequences to about one turn 'up' or 'down'. The comparison is also hampered by the fact that some authors modify the underlying coordinates for bacteriorhodopsin (Rippmann, this article; helix 4 shifted) or rotate or translate some helices slightly to achieve a better fit (Kuipers *et al.*, 1994; Trumpp-Kallmeyer *et al.*, 1992). The handling of the proline induced kinks is another problem which is solved differently by various others. Unfortunately, a critical comparison of the models is not possible, as the coordinates are usually not available.

Prejudice

In the current situation where there is no high resolution structure for a GPCR (which would allow model building by homology of all other GPCRs) one has to rely on a rather large number of 'soft' sometimes ill-founded heterogenous constraints for derivation of three-dimensional models (e.g. prediction of the transmembrane stretches in the sequences, choice of the template structure, mutagenesis data, etc.). The models thus derived usually try to incorporate as many as possible of these constraints, but there are always some which are mutually exclusive, where sometimes an arbitrary decision must be made for or against one set of constraints. For example, should one assume that the Lys residue in helix 7, to which the retinal is attached in bacteriorhodopsin and in rhodopsin is in an equivalent position (possibly conserved by evolution) or should one rely on the transmembrane helical predictions which do not place the functionally equivalent Lys residues in structurally equivalent positions? Should one 'believe' in an evolutionary relationship at all?

The number of potentially useful constraints (e.g. from mutagenesis data) is so large (and fast growing) that it will be important to devise convenient computerized tools for the handling of these data. One attempt in this direction is the accumulation of such information in the publicly accessible 7TM file server at EMBL in Heidelberg (Vriend, 1993–1994).

G protein-coupled receptor models

GPCRs are certainly the pharmacological targets for a great variety of (as yet undiscovered) drugs with enormous importance for public health. Hence the great interest of the pharmaceutical industry in GPCR modelling, even at this early stage where no single method for the derivation of GPCR models has shown a crucial advantage over other approaches. This situation is stimulating for methodological work, comparable to the example of renin modelling which decisively stimulated the development of homology modelling (Raddatz *et al.*, 1983; Sibanda *et al.*, 1984).

Despite their limitations, it is hoped that the above examples show that three-dimensional models of GPCRs can add a new dimension to the study of structure–activity relationships of non-peptidic ligands. Work is underway to evaluate their predictive potential more rigorously. Their value for the study of structure–activity relationships of peptide ligands which are likely to use additional or other binding epitopes outside the transmembrane spanning helices is not yet clear.

Acknowledgements

I would like to thank Julie Thompson and Toby Gibson, EMBL Heidelberg, for various programs to generate the colour coded multiple sequence alignments, and David Jones, University College London, for a modified version of his program MEMSAT. Dan Donnelly and Philippe Cronet kindly provided coordinates of their models.

References

Allen FH, Kennard O, Taylor R. (1983) Systematic analysis of structural data as a research technique in organic chemistry. *Accounts Chem. Res.* **16**: 146–153.

Amit A, Mester L, Klewe B, Furberg S. (1978) The crystal structure of serotonin hydrogen oxalate. *Acta Chem. Scand. Ser. A.* **32**: 267–270.

Bairoch A, Boeckmann B. (1991) The SWISS-PROT protein sequence data bank. *Nucleic Acids Res.* **19**: 2247–2249.

Baker RW, Chothia C, Pauling P, Weber HP. (1972) Molecular structure of LSD. *Science,* **178**: 614–615.

Baldwin JM. (1993) The probable arrangement of the helices in G protein-coupled receptors. *EMBO J.* **12**: 1693–1703.

Barker WC, Hunt LT, George DG, Garavelli JS. (1991) The PIR protein sequence database. *Nucleic Acids Res.* **19**: 2231– 2236.

Benkert O, Gruender G, Wetzel H. (1992) Dopamine autoreceptor agonists in the treatment of schizophrenia and major depression. *Pharmacopsychiat.* **25**: 254–260.

Bernstein FC, Koetzle TF, Williams GJD, Meyer EF, Brice MD, Rodgers JR, Kennard O, Shimanochi T, Tasumi M. (1977) The protein data bank: a computer-based archival file for macromolecular structures. *J. Mol. Biol.* **12**: 535–542.

Boettcher H, Barnickel G, Hausberg H-H, Haase AF, Seyfried CA, Eiermann V. (1992) Synthesis and dopaminergic activity of some 3-(1,2,3,6-tetrahydro-1-pyridylalkyl)-indoles. A novel conformational model to explain structure-activity relationships. *J. Med. Chem.* **35**: 4020–4026.

Clark S. (1991) MALIGNED, a program for display and editing of multiple sequence alignments, University of Toronto; the program is available from the EMBL file server.

Cowan SW, Schirmer T, Rummel G, Steiert M, Ghosh R, Pauptit RA, Jansonius JN, Rosenbusch JP. (1992) Crystal structures explain functional properties of two E. coli porins. *Nature,* **358**: 727–733.

Cronet P, Sander C, Vriend G. (1993) Modeling of transmembrane seven helix bundles. *Prot. Eng.* **6**: 59–64.

Deisenhofer J, Michel H. (1989) Nobel lecture. The photosynthetic reaction centre from the purple bacterium Rhodopseudomonas viridis. *EMBO J.* **8**: 2149–2169.

Donnelly D, Findlay JBC, Blundell TL. (1994) The evolution and structure of aminergic G-protein coupled receptors. *Receptors Channels,* **2**: 61–78.

Engelman DM, Steitz TA, Goldman A. (1986) Identifying nonpolar transbilayer helices in amino acid sequences of membrane proteins. *Ann. Rev. Biophys. Biophys. Chem.* **15**: 321–353.

Fraser CM, Chung F-Z, Wang C-D, Venter JC. (1988) Site directed mutagenesis of human β-adrenergic receptors: Substitution of aspartic acid-130 by asparagine produces a receptor with high-affinity agonist binding that is uncoupled from adenylate cyclase. *Proc. Natl Acad. Sci. USA,* **85**: 5478–5482.

Gribskov M, McLachlan M, Eisenberg D. (1987) Profile analysis: detection of distantly related proteins. *Proc. Natl Acad. Sci. USA,* **84**: 4355–4358.

Heijne von G. (1986) The distrubution of positively charged residues in bacterial inner membrane proteins correlates with the trans-membrane topology. *EMBO J.* **5**: 3021–3027.

Henderson R, Baldwin JM, Ceska TA, Zemlin, F, Beckmann E, Downing KH. (1990) Model for the structure of bacteriorhodopsin based on high-resolution electron cryo-microscopy. *J. Mol. Biol.* **213**: 899.

Hibert M, Trumpp-Kallmeyer S, Bruinvels A, Hoflack J. (1991) Three-dimensional models of neurotransmitter G-binding protein-coupled receptors. *Mol. Pharmacol.* **40**: 8–15.

Jones DT, Taylor WR, Thornton JM. (1994) A model recognition approach to the prediction of all-helical membrane protein structure and topology. *Biochemistry,* **33**: 3038–3049.

Kuipers W, Van Wijngarden I, IJzerman AP. (1994) A model of the 5-HT1A receptor: Agonist and antagonist binding sites. *Drug Design Discovery,* **11**: 231–249.

Lechleiter J, Hellmiss R, Duerson K, Ennulat D, David N, Clapham D, Peralta E. (1990) Distinct sequence elements control the specificity of G protein activation by muscarinic acetylcholine receptor subtypes. *EMBO J.* **9**: 4381–4390.

MaloneyHuss K, Lybrand TP. (1992) Three-dimensional structure for the beta2 adrenergic receptor protein based on computer modelling studies. *J. Mol. Biol.* **225**: 859–871.

Nordvall G, Hacksell U. (1993) Binding-site modeling of the muscarinic m1 receptor: A combination of homology based and indirect approaches. *J. Med. Chem.* **36**: 967–976.

Oliveira L, Paiva ACM, Vriend G. (1993a) A common motif in G protein coupled seven transmembrane helix receptors. *J. Comput.- Aided Mol. Design,* **7(6)**: 649–658.

Oliveira L, Paiva ACM, Sander C, Vriend G. (1993b) A model for G protein interaction in G protein coupled receptors. *7TM, pre-issue* **2** (Available from vriend@embl-heidelberg.de).

Pearson W.R. (1990) Rapid and sensitive sequence comparison with FASTP and FASTA. *Methods Enzymol.* **188**: 63–98.

Persson B, Argos P. (1994) Prediction of transmembrane segments in proteins utilising multiple sequence alignments. *J. Mol. Biol.* **237**: 182–192.

Raddatz P, Schittenhelm C, Barnickel G. (1983) Computer-graphische Methoden in der Pharmaforschung: Veranschaulichung von Renin-Inhibitor-Komplexen. *Kontakte,* **3**: 3–13.

Rippmann F. (1994) Molecular modeling of G protein-coupled receptors: The ligand gives the clue. 7TM, pre-issue **4**. (available from vriend@embl-heidelberg.de).

Rippmann F, and Boettcher H. (1993) Molecular modelling of serotonin receptors. *7TM, pre-issue* **1** (available from vriend@embl-heidelberg.de).

Roeper D, Jacoby E, Krueger P, Engels M, Groetzinger J, Wollmer A, Strassburger W. (1994) Modeling of G-protein coupled receptors with bacteriorhodopsin as a template. A novel approach based on interaction energy differences. *J. Rec. Res.* **14**: 167–186.

Saltzman AG, Morse B, Whitman MM, Ivanshchenko, Y Jaye, M Felder S. (1991) Cloning of the human serotonin 5-HT2 and 5-HT1C receptor subtypes. *Biochem. Biophys. Res. Commun.* **181**: 1469–1478.

Savarese TM, Fraser CM. (1992) In vitro mutagenesis and the search for structure-function relationships among G protein-coupled receptors. *Biochem. J.* **283**: 1–19.

Seyfried CA, Greiner HE, Haase AF. (1989) Biochemical and functional studies on EMD49980: a potent, selectively presynaptic D-2 dopamine agonist with actions on serotonin systems. *Eur. J. Pharmacol.* **160**: 31–41.

Sibanda B, Blundell T, Hobart PM, Fogliano M, Bindra JS, Dominy BW. Chirgwin JM. (1984) Computer graphics modelling of human renin. *FEBS Lett.* **174**: 102–111.

Smith S. (1990) GDE (Genetic Data Environment) revision 2.0, available via Gopher from megasun.bch.umontreal.ca.

Strader CD, Sigal IS, Dixon RAF. (1989) Structural basis of β-adrenergic receptor function. *FASEB J.* **3**: 1825–1832.

Swindells MB, Thornton JM. (1991) Modelling by homology. *Curr. Opin. Struct. Biol.* **1**: 219–223.

Taylor WR. (1987) Multiple sequence alignment by a pairwise algorithm. *Comput. Appl. Biosci.* **3(2)**: 81–7.

Taylor WR, Jones DT, Green NM. (1994) A method for alpha-helical integral membrane protein fold prediction. *Proteins,* **18**: 281–294.

Trumpp-Kallmeyer S, Hoflack J, Bruinvels A, Hibert M. (1992) Modeling of G-protein coupled receptors: Application to dopamine, adrenaline, serotonin, acetylcholin, mammalian opsin receptors. *J. Med. Chem.* **35**: 3448-3462.

Vriend G. (1990) WHAT IF: A molecular modeling and drug design program. *J. Mol. Graphics,* **8**: 52–56.
Vriend G. (1993–1994) The 7TM file server. *7TM, pre-issues* **1–4** (1993–1994) (available from vriend@embl- heidelberg.de).
Westkaemper RB, Glennon RA. (1993) Molecular graphics models of members of the 5-HT2 subfamily: 5-HT2A, 5-HT2B, 5-HT2C receptors. *Med. Chem. Res.* **3**: 317–334.
Yamamoto Y, Kamiya K, Terao S. (1993) Modeling of human thromboxane A2 receptor and analysis of the receptor-ligand interaction. *J. Med. Chem.* **36**: 820–825.
Zifa E, Fillion G. (1992) 5-Hydroxytryptamine receptors. *Pharmacol. Rev.* **44**: 401–458.

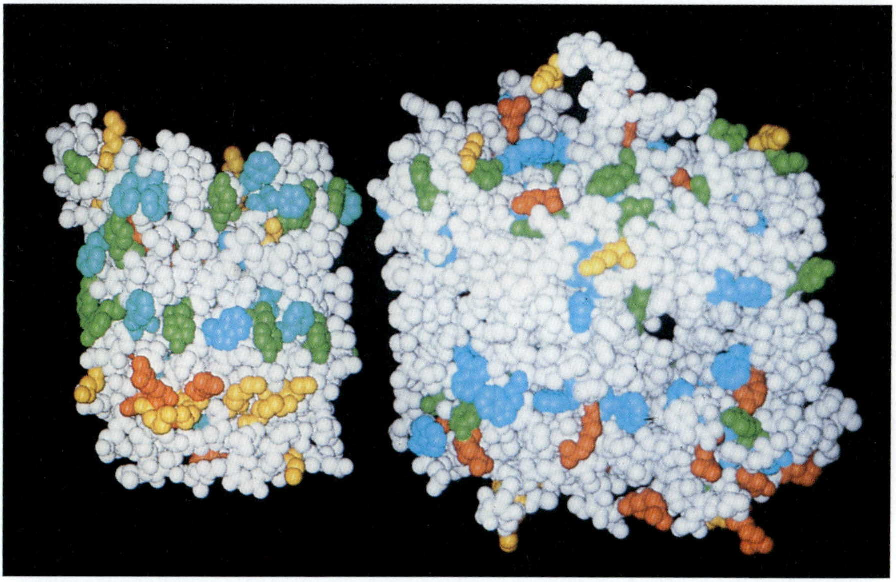

Chapter 6, Figure 1. Photosynthetic reaction centre (larger structure) and porin as examples of crystallographically determined integral membrane protein structures. Trp and Tyr (and Phe in porin) are blue/greenish, Lys and Arg are orange and red. In the porin structure, the characteristic 'stripes' of aromatic polar residues (which are thought to interact with the polar phospholipid headgroups from within the membrane) show up clearly. The two structures also indicate the difference in the thickness of the different membrane types. Lys and Arg residues in porin are found predominantly on the inner surface of the membrane ('positive inside rule').

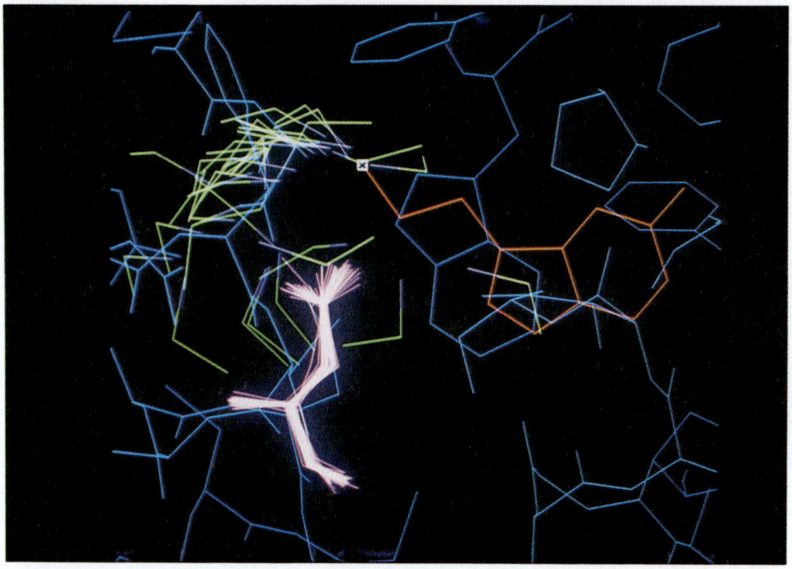

Chapter 6, Figure 4. Interaction of an Asp side chain in a helical context (magenta) with a Lys side chain (any context, green). This interaction was used to find a suitable position for the charged nitrogen of serotonin (red) relative to the crucial Asp on helix 3.

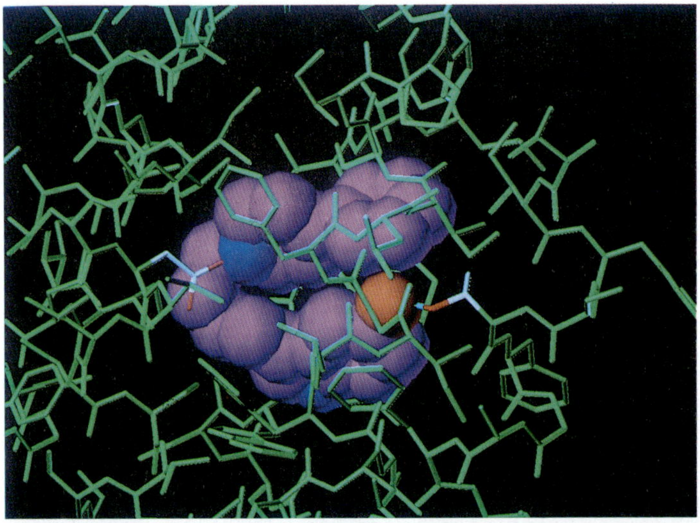

Chapter 6, Figure 5. Roxindole (space filling representation) in the 5-HT_{1A} model (stick representation). In this conformation, the molecule occupies both the 'usual' binding site for small ligands (e.g. serotonin between helices 3, 4, 5, 6 (right)) and some volume on top of it, making various hydrophobic and aromatic interactions. The crucial Asp (white) in helix 3 which may interact with the charged nitrogen of the ligands is on the left; the Thr residue (white) in helix 5 putatively donating a hydrogen bond to the phenolic hydroxy group is shown on the right.

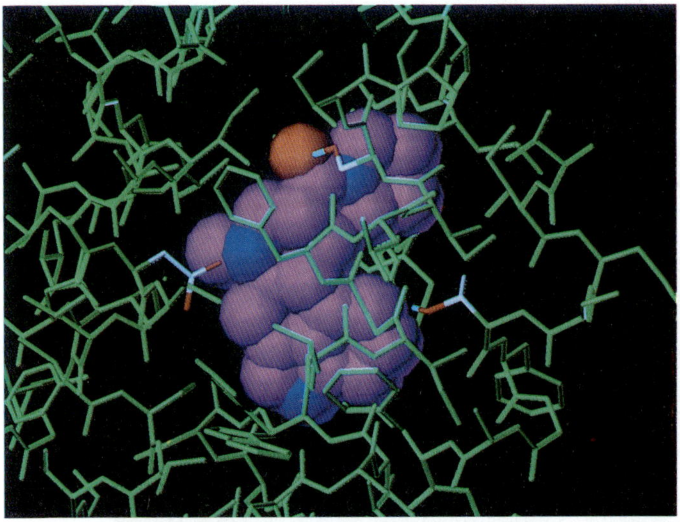

Chapter 6, Figure 6. *d*-LSD (space filling representation) in the 5-HT_{1A} model (stick representation). The peptide bond (top) accepts a stereospecific hydrogen bond from a Ser residue on helix 4. This could explain why only the *d*-isomer is active.

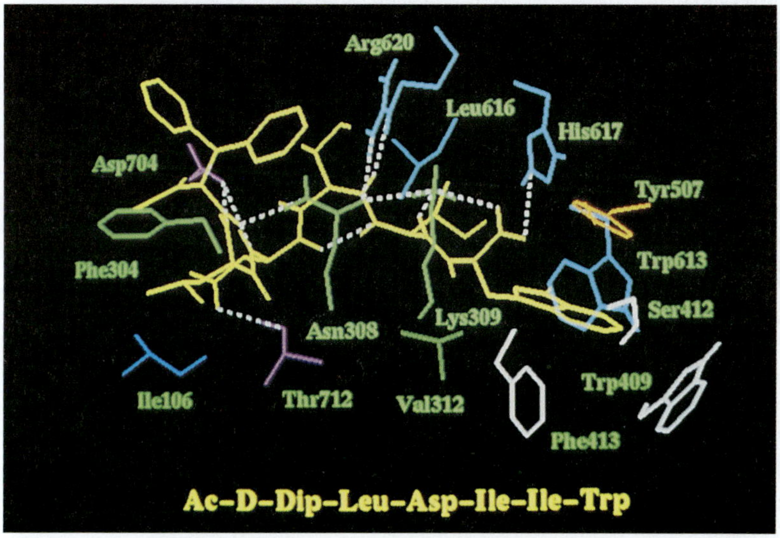

Chapter 7, Figure 10. Details of the interaction between the functional antagonist PD 142893 (compound 9, *Table 1*) and the ETA binding site residues, as resulted from quenched MD minimization.

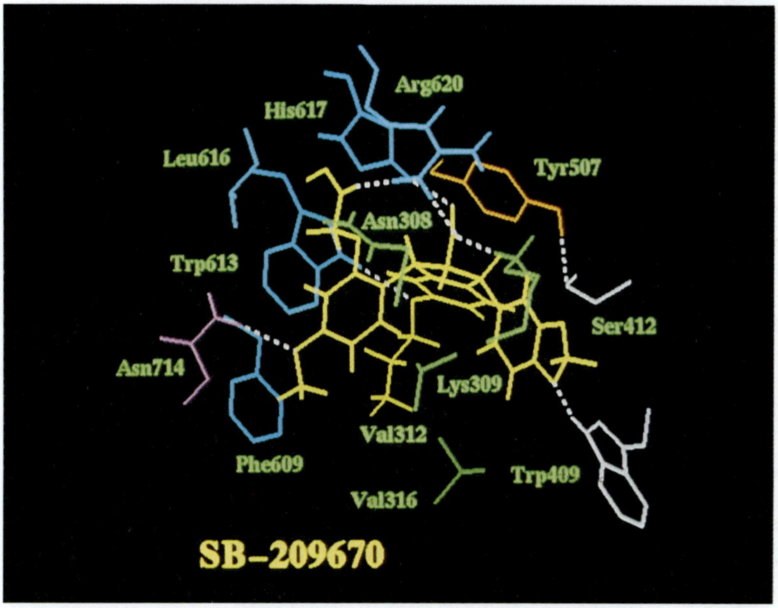

Chapter 7, Figure 11. Details of the interaction between the functional antagonist SB 209670 (compound 1, *Table 1*) and the ETA binding site residues, as resulted from quenched MD minimization.

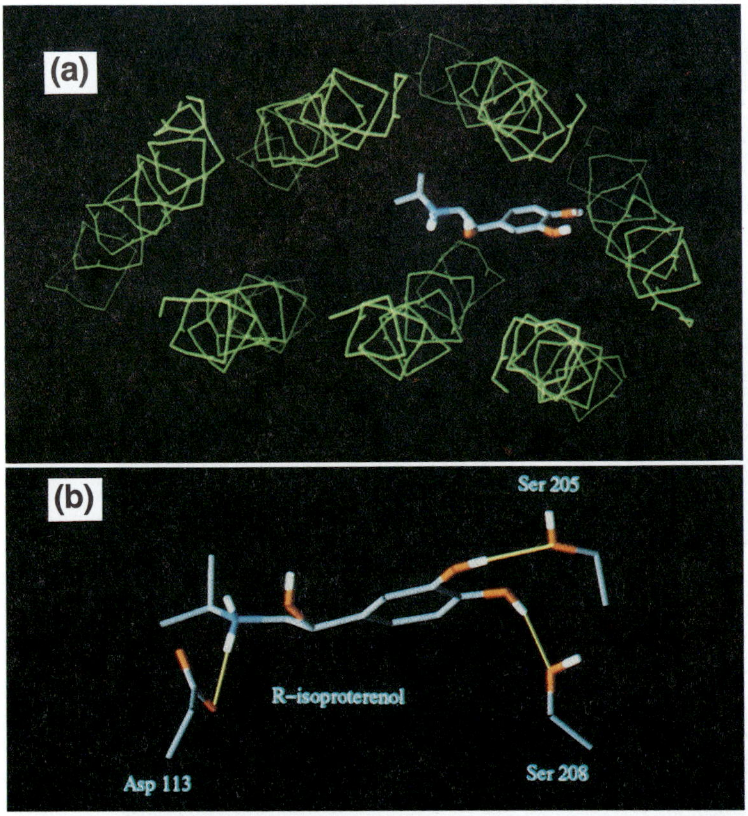

Chapter 8, Figure 4. (a) The binding of isoprenaline (grey: carbon; blue: nitrogen; red: oxygen; white: hydrogen atoms connected to oxygen or nitrogen) to the human β_2-adrenoceptor (direction of view from outside to inside the cell). The receptor backbone (green) only is shown, without side chains.
(b) The anchoring of isoprenaline by three residues in the human β_2-adrenoceptor (see text). Colouring as in *Figure 4(a)*. Yellow lines indicate possible hydrogen bonds. Note that $Ser^{205,208}$ should read $Ser^{204,207}$.

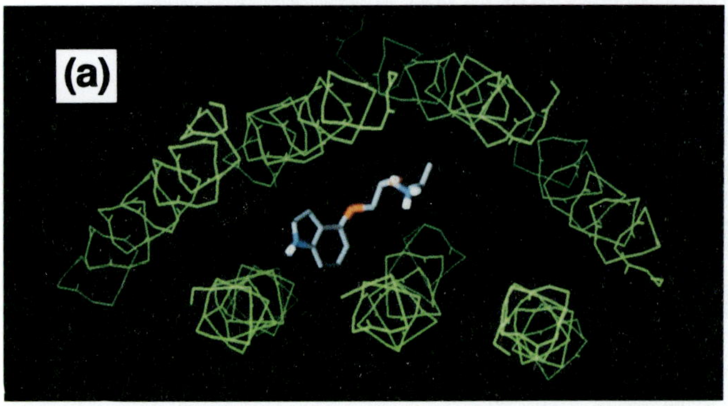

Chapter 8, Figure 5. (a) The binding of pindolol (grey: carbon; blue: nitrogen; red: oxygen; white: hydrogen atoms connected to oxygen or nitrogen) to the human β_2-adrenoceptor (direction of view from outside to inside the cell). The receptor backbone (green) only is shown, without side chains.

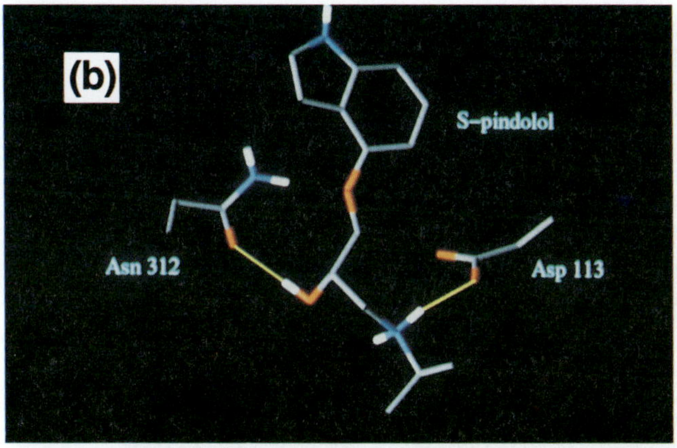

Chapter 8, Figure 5. (b) The anchoring of pindolol by two residues in the human β_2-adrenoceptor (see text). Colouring as in *Figure 5(a)*. Yellow lines indicate possible hydrogen bonds.

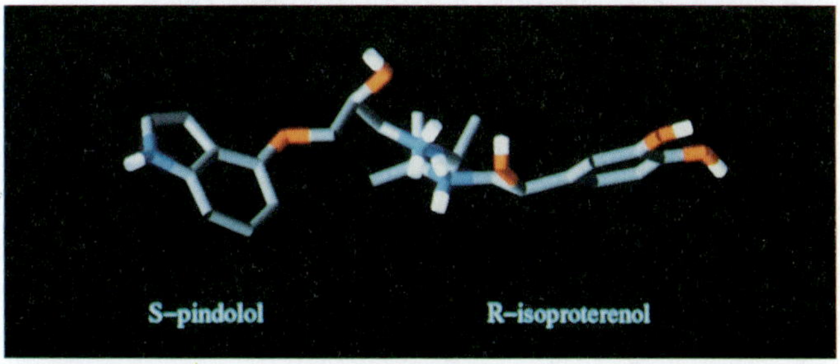

Chapter 8, Figure 6. S-pindolol and R-isoprenaline in their receptor-bound conformations.

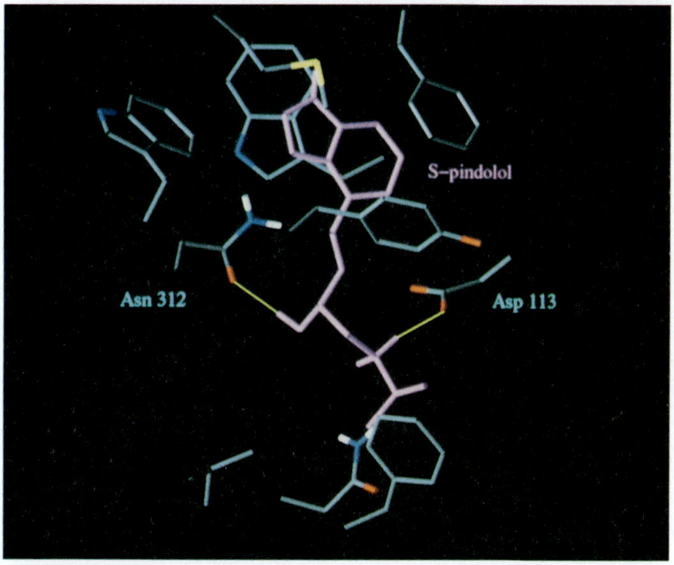

Chapter 8, Figure 7. The pindolol binding site on the human β_2-adrenoceptor. All amino acids within 4 Å of the ligand are shown. Pindolol is in pink (yellow: sulphur). Yellow lines indicate possible hydrogen bonds.

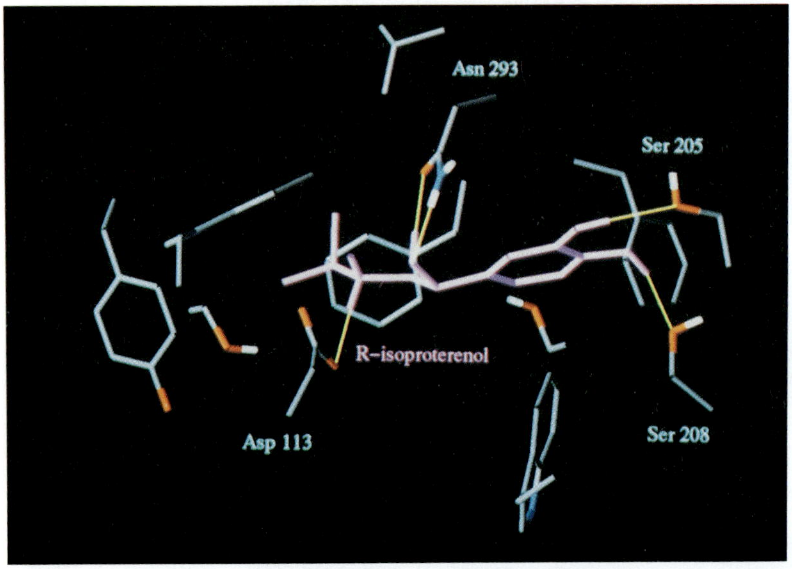

Chapter 8, Figure 8. The isoprenaline binding site on the human β_2-adrenoceptor. All amino acids within 4 Å of the ligand are shown. Isoprenaline is in pink. Yellow lines indicate possible hydogen bonds. Note that $Ser^{205,208}$ should read $Ser^{204,207}$.

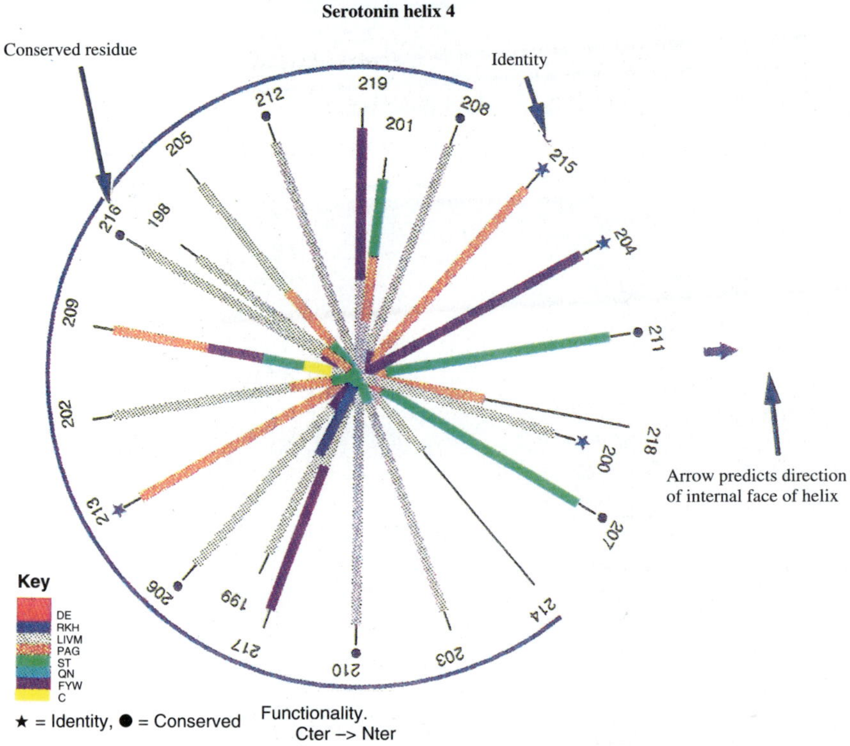

Chapter 10, Figure 6. A typical *Helanal* wheel with surrounding hydrophobic arc in blue. This shows that helix 4, being mostly hydrophobic, is mainly exposed to the membrane. Key shows groups and associated colours.

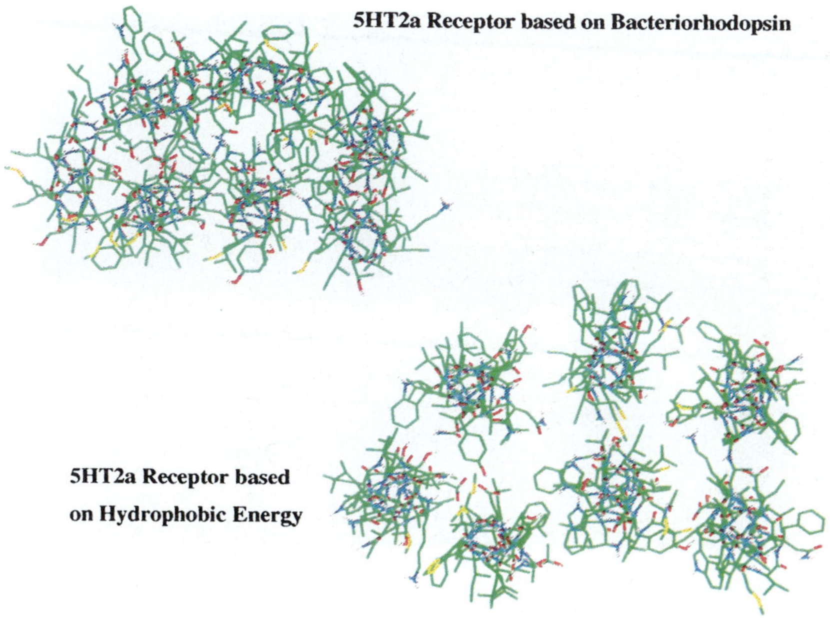

Chapter 10, Figure 7. (Top) a model of 5-HT_{2a} based on bacteriorhodopsin. (Below) an early 5-HT_{2a} model based on maximization of hydrophobic energy.

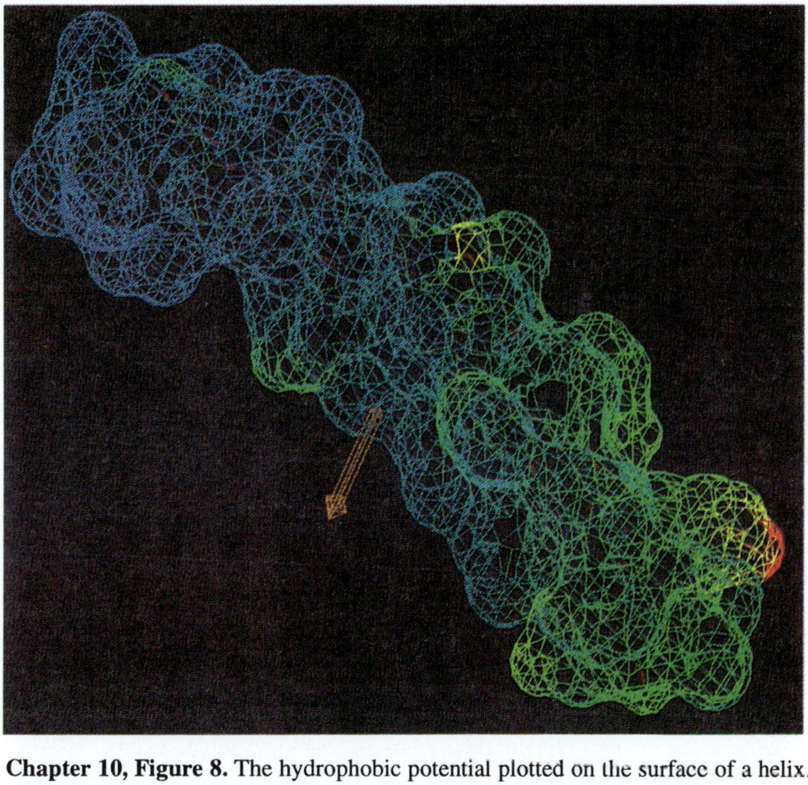

Chapter 10, Figure 8. The hydrophobic potential plotted on the surface of a helix. The three-dimensional arrow is the Donnelly conservation moment.

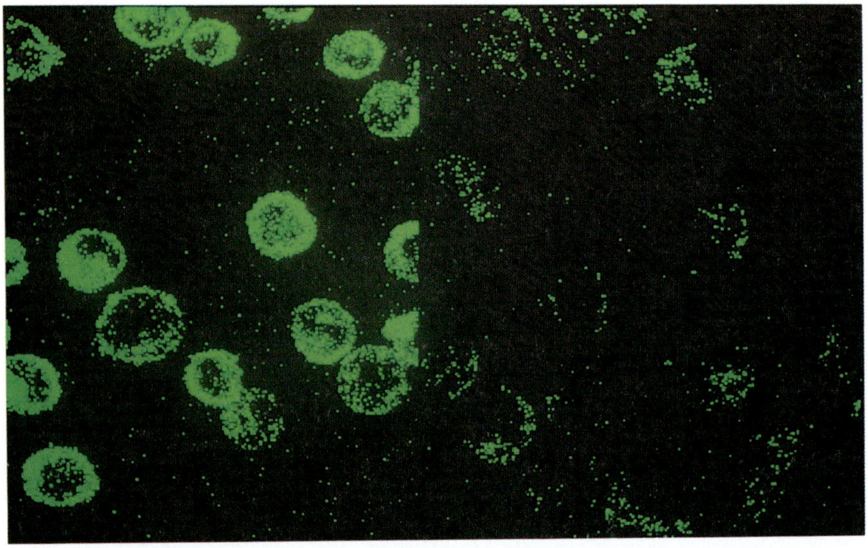

Chapter 11, Figure 2. Confocal microscopy analysis of DAB-N binding to NK2 receptors on CHO cells (excitation at 488 nm, emission at 510–525 nm; magnification 720×). Left panel: 1.4 nM DAB-N; right panel: 1.4 nM DAB-N and 20 μM GR94800.

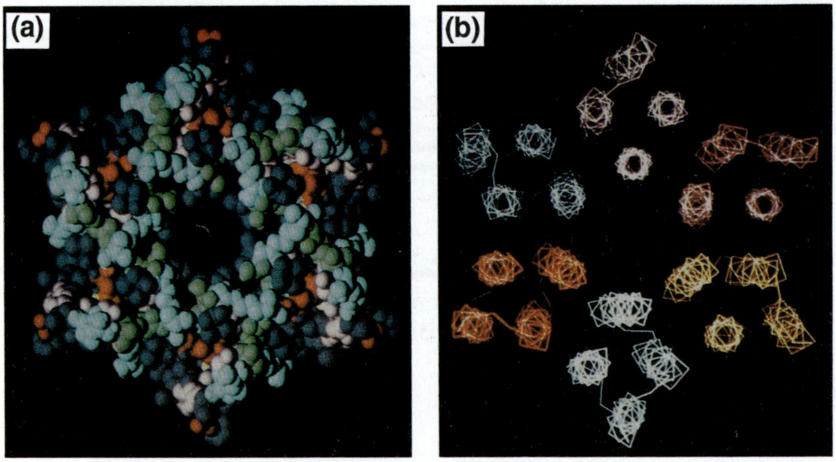

Chapter 12, Figure 3. Computergraphic modelling of the 16 kDa proteolipid The three-dimensional model, constructed as described in the text, is viewed perpendicular to the plane of the membrane. (a) Space-filling model. (b) Main chain atoms only. Reproduced from Finbow *et al.* (1992) by permission of Oxford University Press.

Chapter 7

The heuristic-direct approach to quantitative structure–activity relationship analysis of ligand–G protein-coupled receptor complexes

M. Cristina Menziani, Francesca Fanelli, Marina Cocchi and Pier G. De Benedetti

1. Introduction

Theoretical quantitative structure–activity relationship (QSAR) analysis is now a well-established tool which employs a large variety of theoretical molecular descriptors derived both from quantum mechanical calculations and computer assisted molecular modelling procedures, in order to rationalize the variation of the common bio-activity within an appropriate and informative set of molecules. This implies that the bio-activity is reduced and translated into the chemical formalism, through the maximization of the molecular complementarity between the ligand and the target of unknown three-dimensional structure (indirect approach or ligand fitting) or known three-dimensional structure (direct approach or receptor fitting), as outlined in *Figure 1* (Menziani and De Benedetti, 1991a).

Until recently, the qualitative potential of the direct approach has been successfully exploited in drug design and structure–reactivity relationship (SAR) studies of soluble enzymes, DNA and carrier proteins (Greer *et al.*, 1994). Nowadays its application can be extended to other systems, such as membrane-bound receptor proteins, since a valuable alternative to direct structure determination has been proposed by protein homology modelling strategies.

The explosive growth of information about the molecular pharmacology of receptor-triggered processes, together with the cloning, sequencing and expression of a variety of membrane-bound receptors provide an added dimension for

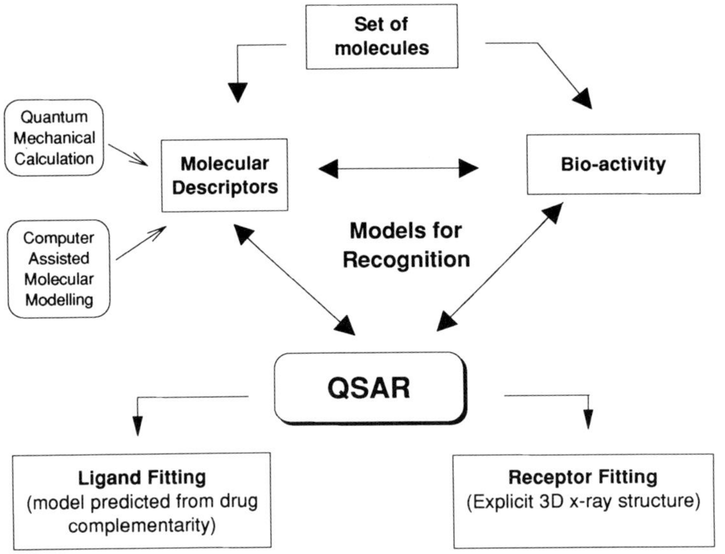

Figure 1. The theoretical quantitative structure–activity relationship (QSAR) approach to drug design.

structure–activity studies, at the level of the receptor (Hollenberg, 1990); from these studies a detailed picture of the molecular basis of receptor function emerges. Knowledge-based modelling techniques can be used to ascribe the deduced function (i.e. the observed properties) to a certain three-dimensional structure. In fact, in the absence of direct crystallographic data, reasonable three-dimensional models have been elaborated for several members the G protein-coupled receptor (GPCR) superfamily by the synergistic exploitation of heuristic prediction and experimental data (for a review, see Ballesteros and Weinstein, 1994).

Moreover, molecular simulations of ligand–receptor interactions provide intimate details of the intermolecular complementarity mutually achieved upon binding, and allow the inference of mechanistic conclusions regarding the role of the different structural components. However, at the current time the vast majority of the studies on ligand–GPCR interactions are primarily qualitative (SAR).

A quantitative extension has been presented recently with the heuristic-direct QSAR approach (De Benedetti *et al.*, 1993). This approach is able to handle the heterogeneous experimental information about the ligands and targets and synthesize and translate them into quantitative models. To set complex molecular modelling experiments in the conceptual framework of correlation analysis has a two-fold goal, namely (a) to challenge the ability of the three-dimensional models to agree with the pharmacological data in quantitative details and (b) to consider the drastic assumptions and approximations inherent in the methodology used more acceptable; as a consequence, any effort can be focused on improving the predictive and interpretative aspects of the model.

Relevant QSAR studies can only be obtained by using chemical descriptors which possess high information content and are easily interpretable. For this purpose, molecular simulations of ligand–receptor interactions allow the definition of intermolecular interaction descriptors, *ad hoc* for the problem under study, which help rationalize, through the formulation of statistical models, the biological responses (affinity, selectivity, efficacy) (Fanelli *et al.*, 1994a, 1994b; Menziani *et al.*, 1989; Menziani and De Benedetti, 1991b; Zhang and Weinstein, 1993a). In this way, mechanistic interpretations inferred by SAR studies can be corroborated or refuted, on quantitative grounds. Moreover, new lead compounds can be designed on the basis of the intermolecular interaction analysis and new protocols for mutagenesis or biophysical experiments can be drawn up, in order to improve the predictive and interpretative aspects of the models. *Figure 2* shows an outline of the quantitative approach to ligand–receptor interactions.

2. Application: the endothelin receptors

A manifold example of the implementation of the heuristic-direct approach to QSAR analysis of ligand–receptor complexes is reported here for the interaction of the endothelin ETA and ETB receptor subtypes with two series of peptide and non-peptide antagonists.

The endothelins (ET-1, ET-2 and ET-3) are potent and long acting vasoconstrictor peptides which act on vascular and non-vascular smooth muscle. Their sequence consists of 21 amino acids. The C-terminal portion is shared by all of the members of the mammalian endothelin family; the amino acid substitutions between the isopeptides are clustered in the amino-terminal loop portions, especially within position 2–7, as showed in *Figure 3*. The wide variety of diverse actions exerted by endothelins are mediated by specific receptors with discrete and regulated cellular distribution and function (Huggins *et al.*, 1993a). They belong to the superfamily of heptahelical GPCRs and the current classification is based on their relative affinities for the three ET isoforms: the ETA receptor is ET-1 selective, showing an affinity rank order of ET-1=ET-2»ET-3, whereas the ETB receptor exhibits similar affinities for all three isopeptides. In addition, the existence of a third type ETC, with highest affinity for ET-3, has only been demonstrated, so far, in non-mammalian cells (Masaki *et al.*, 1994).

Little is known about the receptor region involved in ligand binding. Preliminary evidence, resulting from chimeric receptors, indicates that there might be at least two different ligand interaction subdomains within the ET receptors: residues located at the entrance of the binding cleft seem to bind to the N-terminal loop of

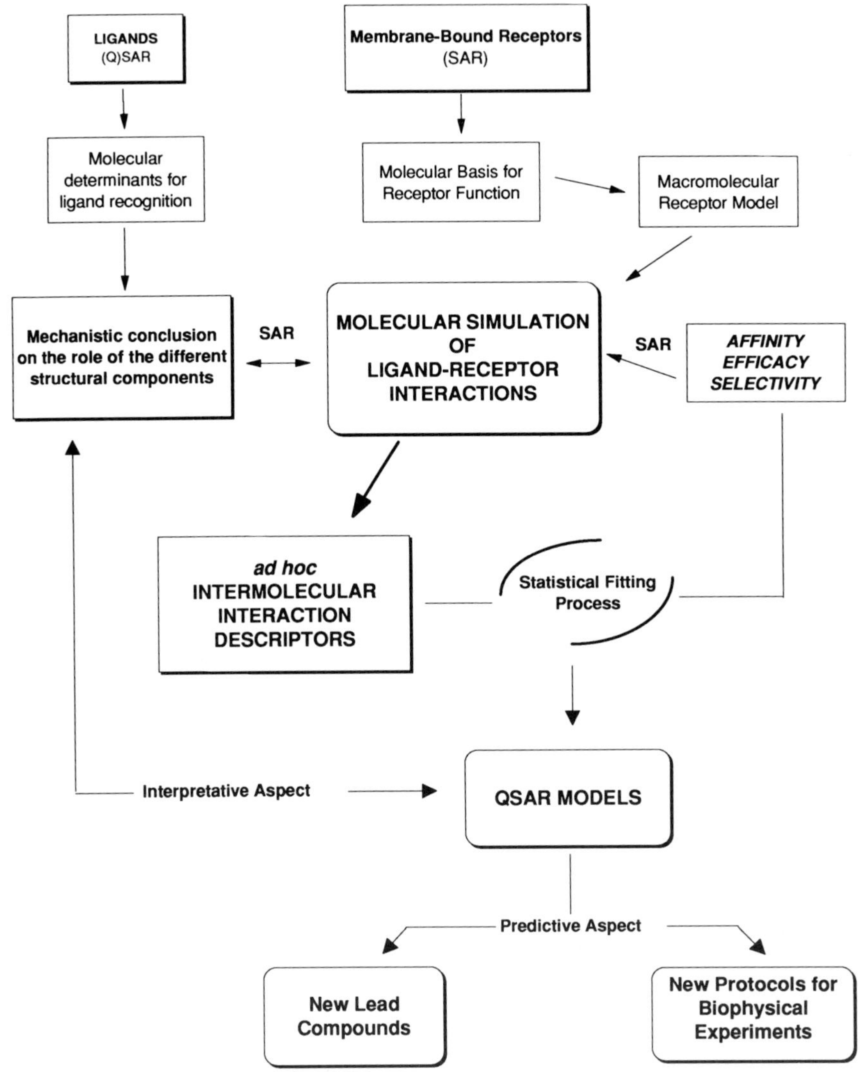

Figure 2. Flow chart for the quantitative approach to ligand–receptor interaction.

ET-1 (but not that of ET-3), which confers high affinity and selectivity to the endogenous peptides, while the transmembrane region of the receptors seems to bind to the C-terminal part of the endogenous ligands, which is known to be highly important for receptor recognition and signal transduction. In particular, the N-terminal tail and the first extracellular loop of both the receptor subtypes with their flanking transmembrane domains (H1, H2 and H3) are crucial for the interaction with ET-1; critical binding determinants for ETB-selective antagonists

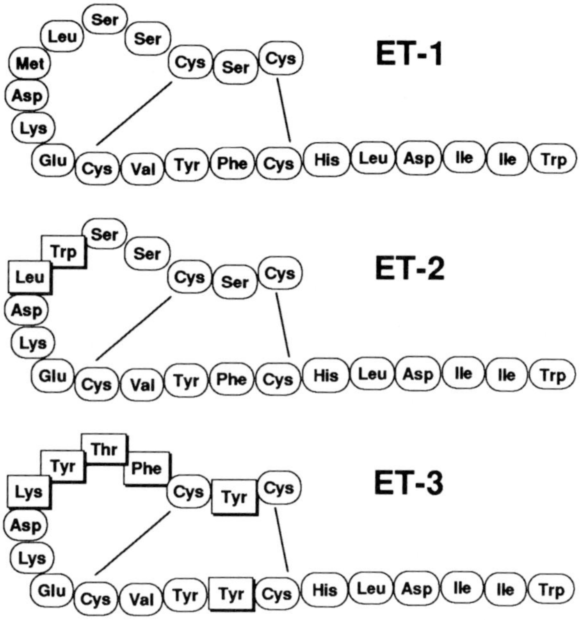

Figure 3. The primary sequences of the endogenous endothelin ligands, showing the positions of the disulphide bonds. Residues which differ from those of endothelin-1 are in square boxes.

appear to reside in portion of the sequence between the second and third extracellular loops (including transmembrane domains H4–H6); selectivity determinants for the synthetic antagonist BQ123 seem to be also located in the region including transmembrane domain H1–H2 and the first extracellular loop, however, the involvement of the third extracellular loop and of transmembrane domain H7 cannot be excluded (Adachi *et al.*, 1994; Becker *et al.*, 1994; Sakamoto *et al.*, 1993).

The role of specific residues of the transmembrane domains in the binding of peptide and non-peptide ligands has been investigated by site-directed mutagenesis studies. Replacement of Lys-181 (H3) of rat ETB by Asp reduces receptor affinity for ET-3 and sarafotoxin S6C, with no loss in transduction ability (Zhu *et al.*, 1992). Controversial results very recently reported on the role of Tyr-129, in H2 of the ETA receptor (Krystek *et al.*, 1994; Lee *et al.*, 1994) emphasize the molecular complexity of ligand–endothelin receptor interactions and suggest that the ligand selectivity for the ETA receptor subtype may involve multiple receptor regions that act in synergy. Moreover, chemical modifications of histidine residues (Huggins *et al.*, 1993b) and tyrosine residues (Bousso *et al.*, 1992) produce inhibition of the binding of the ligands to the ET receptors, however, the location of these critical residues is not clear.

3. Computational strategy

The general strategy adopted in this study is outlined in *Figure 4*.

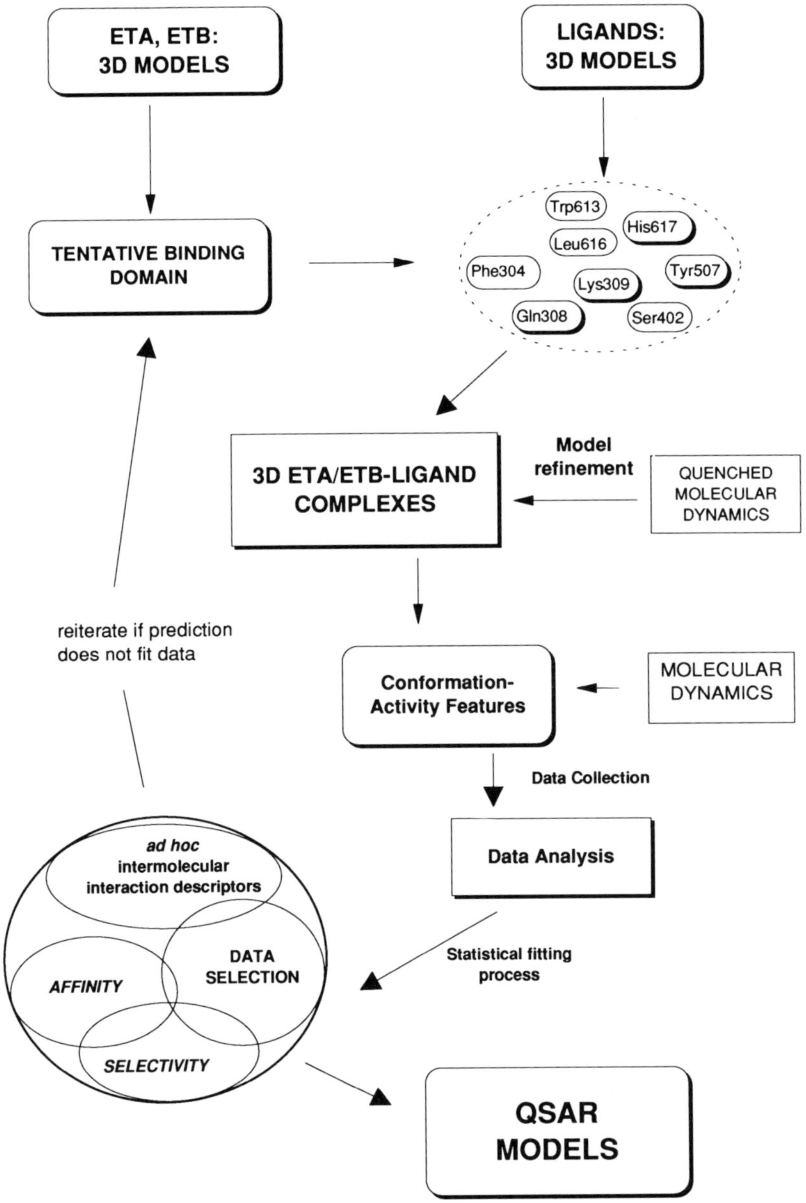

Figure 4. Computational strategy adopted in the heuristic-direct QSAR approach to ligand–receptor interactions.

Three-dimensional models of the ETA and ETB receptors

The ETA and ETB transmembrane domains were modelled using methods previously described (Ballesteros and Weinstein, 1994; Fanelli *et al*, 1994a; Fanelli *et al.*, 1995; Trumpp-Kallmeyer *et al.*, 1992). Extracellular loops were added using the secondary structure prediction methods implemented in the Quanta program and molecular dynamics (MD) with weak harmonic constraints.

Tentative binding domain

The antagonists were manually docked in a tentative binding domain which conforms to the experimental information available on the ET receptor subtypes and adds a few residues which were shown to be important in other peptide receptors and in the cationic neurotransmitter and mammalian opsin receptors (Trumpp-Kallmeyer *et al.*, 1992). The cardinal point of the docking procedure is the formation of a charge-reinforced hydrogen bond between Lys-309 (the first digit corresponds to the helix, the next two digits indicate the position of the residue in the helix) and the carboxylic head of the antagonists considered. A Lys residue in this position is only present and conserved in the ET receptors. Other proposed key binding site residues are: Gln-308, His-617, Ser-402, Phe-304, Tyr-507, Trp-613 and Leu-616. Gln-308 occupies the same position as the Asp conserved in cationic neurotransmitter receptors which has been proposed to form an ionic interaction with the cationic head group of their ligands (Savarese and Fraser, 1992). His-617, mainly conserved in peptide receptors, seems to be involved in the binding of certain non-peptide antagonists of the neurokinin-1 receptor, but not in that of the natural peptide agonist (Fong *et al.*, 1994; Zoffmann *et al.*, 1993). Ser-402 is proposed to make an hydrogen bond with the indole amine of the C-terminal Trp of endothelin peptides on the basis of the analogy of this Trp with the cationic neurotransmitter serotonin (Huggins *et al.*, 1993b). Finally, Phe-304, Tyr-507, Trp-613, and Leu-616 delimit the hydrophobic pocket which enhances the strength of the ionic interaction between Lys-309 and the ligands. Trp-613 is conserved throughout the GPCR superfamily, while Phe-304 and Leu-616 are substituted for Trp and Phe, respectively, in the cationic neurotransmitter receptors. Tyr-507 is only present and conserved in the ET receptor subtypes.

Model refinement

The minimized coordinates of the complexes were used as starting points for two series of simulations, which combine a sequence of quenched molecular dynamics (QMD) simulations, and a 80 picoseconds molecular dynamics (MD) run at 300 K. Weak harmonic constraints (30 $kJmol^{-1}$ $Å^{-1}$) were applied by using the NOE facility in the CHARMm program in order to maintain the helical structure during dynamics simulation. The docking of peptide ligands is complicated by several difficulties, (i.e. their conformational flexibility and the possibility of multiple sites of interaction, which can easily lead to biased local minima when traditional energy minimization alone is used). Unfavourable interactions that remain after energy minimization can be removed to a limited extent, by using QMD, which allows extension of the conformational space search. Moreover, the structural/dynamic features of the ligands and of the complexes need to be investigated using MD techniques in order to evaluate the molecular flexibility requirements for achieving optimum intermolecular complementarity and in an attempt to decipher the mechanism(s) of information transfer upon binding (Menziani *et al.*, 1991c; 1993; Zhang and Weinstein, 1993b).

Data analysis

To ensure consistency in the interaction modes of the ligands considered, an extensive data analysis has to be carried out. Moreover, the active site residues which play a fundamental role in the binding of the antagonists, or in establishing inter-helix interaction networks can be selected on the basis of the frequency of occurrence of these specific interactions during the simulation period of the dynamic run.

Data selection

Statistical procedures provide a tool for the selection of the most informative molecular descriptors to predict and explain the biological response (referred to as *ad hoc* intermolecular interaction descriptors, in *Figure 4*). However, the application of these methods, particularly those which make use of the statistical multivariate approach in the data analysis, is mainly devoted to improving the predictive aspect of QSAR analysis; this is usually achieved at the expense of the physical interpretability of the QSAR models produced. We deem that the fundamental importance in drug design of the interpretative aspects of QSAR modelling must be emphasized, hence our primary goal is the search for a linear QSAR model with meaningful descriptors. Thus, when a correlation is found, we know how to interpret it. The search for a linear QSAR model is also justified in the conceptual framework of the physical organic chemistry, both on the grounds of its local validity (Sjostrom and Wold, 1981) and in the kinetic description of the rate determining step of drug–receptor interaction (Richards, 1983).

Notwithstanding the highly speculative nature of the models obtained by this strategy and their local unreliability, they can be regarded as useful tools for the interpretation of experimental results, for suggesting trends and correlations and for facilitating the design of ligands with improved affinity and selectivity. As new experimental data emerge, the models will hopefully be modified, corrected, improved and refined to fit the new data, thus enhancing their validity.

4. Results and discussion

Peptide ETA and ETB receptor antagonists

Structure–activity studies of the C-terminal hexapeptide of endothelin have been carried out recently (Doherty *et al.*, 1993) in order to elucidate the amino acids important for receptor binding and for agonist or antagonist activity. ET(16–21) has been shown to be the minimum sequence with weak binding affinity at both the ETA and ETB receptor subtypes (50–70 μM) and partial agonist activity in the guinea-pig isolated bronchus. Modification of the hexapeptide has led to a series of compounds with enhanced binding affinity and antagonistic behaviour.

With the aim of evaluating the conformational, topographical and dynamic features important for recognition and of elucidating quantitatively the receptor structure–affinity/selectivity relationships, we have focused our attention on the series of C-terminal endothelin hexapeptide antagonists, reported in *Table 1*. The experimental binding affinities (Doherty *et al.*, 1993) and the computed interaction energies with respect to the ETA and ETB receptor subtypes (Menziani *et al.*, 1995a) are also listed in *Table 1* for the selected peptide antagonists.

The quantitative relationships obtained between the binding affinity data values (pIC_{50}) and the computed interaction energies (IE) are shown in *Figure 5(a)* and *(b)*. Because of their statistical soundness, these correlations provide us with a precious tool to understand and rationalize the experimental data and to guide the design of new compounds with improved characteristics. Moreover, a convincing example of theoretical quantitative structure-selectivity relationships is given by correlating $\Delta pIC5_{50}$ (which represents the selectivity of a given compound for the two receptor subtypes examined) and ΔIE (the difference in the computed interaction energy of a given compound with the two receptors subtypes) (see *Figure 5(c)*).

Data analysis of the QMD minimized complexes shows that ligand–receptor interaction occurs mainly within the transmembrane domain of the receptor and involves, to differing extents, all the seven helices in both the ETA and ETB receptor subtypes. The antagonists which carry diphenylalanine (Dip) at the N-terminal position of their sequence interact with a few residues in the extracellular loops; however, the incidence of these interactions on the total interaction energy of the ligands is less than 1%.

The transmembrane amino acid residues of the ETA and ETB receptor subtypes which are involved in the interaction with the 12 peptide antagonists are listed in *Figure 6*, where an average situation of the general binding preferences is outlined. However, the receptor flexibility is not significantly hindered by the intermolecular interactions within the ligand binding domain, hence the role of molecular flexibility in ligand binding has to be considered explicitly for a deeper understanding of the recognition phenomena.

Table 1. Experimental binding affinities (pIC_{50}) and computed interaction energies (IE, kcal/mol) for the C-terminal endothelin hexapeptide antagonists complexed with the ETA and ETB receptor subtypes

				ETA		ETB	
N	Antagonists			PIC_{50}	IE	PIC_{50}	IE
1	Ac-D-His	Leu	Asp Ile Ile Trp	5.05	−130.13	5.04	−127.36
2	Ac-D-Phe	Leu	Asp Ile Ile Trp	5.74	−147.45	5.70	−140.96
3	Ac-Phe	Leu	Asp Ile Ile Trp	4.73	−132.93	4.73	−127.64
4	Ac-D-Phe	Phe	Asp Ile Ile Trp	6.70	−154.47	7.22	−156.29
5	Ac-D-Phe	D-Phe	Asp Ile Ile Trp	6.04	−148.27	7.70	−157.14
6	Ac-D-Phe	Leu	Phe Ile Ile Trp	5.85	−147.25	7.44	−150.89
7	Ac-D-Phe	Orn	Asp Ile Ile Trp	6.23	−162.72	5.61	−134.40
8	Ac-D-Phe	Leu	Asp Ile Ile Tyr	4.70	−126.91	6.82	−144.10
9	Ac-D-Dip*	Leu	Asp Ile Ile Trp	7.40	−173.65	7.22	−156.58
10	Ac-D-Dip	Leu	he Ile Ile Trp	6.24	−155.42	7.22	−153.00
11	Ac-D-Dip	Orn	Asp Ile Ile Trp	7.90	−187.03	6.82	−142.61
12	Ac-D-Dip	Leu	Asp Ile Ile Tyr	5.57	−150.13	5.21	−138.75

*Dip=diphenylalanine.

The structural/dynamics properties which are more representative of the differences between the ligand–ETA and ligand–ETB receptor complexes can be captured by the analysis of the simulation period of the dynamics run. The following important observations arise: (a) in the manual docking step His-617 was positioned in the proximity of the negative carboxyl-terminus of the interacting ligands, for both receptor subtypes. During dynamics it establishes alternatively a charge reinforced hydrogen bond with the C-terminus or with the peptide bond of the ligands interacting with the ETA receptor. In contrast, in the complexes between the ETB receptor subtype and ligands 1, 3, 5, 8 and 12 in *Table 1*, it slowly moves away, during the warm up phase, and finds a more favourable environment in the neighbouring helix 5. (b) The antagonists interfere to differing extents, with a large network of hydrogen bonds, which connects H2, H3 and H7 in the two isolated receptor subtypes. As shown in *Figure 7,* this network involves Ser-311-Asp-211-Ser-715 and Thr-315-Asp-211-Asn-718 in both receptor subtypes and can be extended to Tyr-214-Ser-310-Asp-211 in the ETA receptor. (c) Ser-412 and Trp-409 are dynamically competitive in establishing interactions with the C-terminal amino acids of the ligands. (d) Arg-620 forms an ionic pair with Asp-704 in both receptor subtypes (see *Figure 7*). Only those ligands which have the diphenylalanine moiety as their first residue are able to break this strong inter-helix interaction by improving their interaction energy by means of charge-reinforced hydrogen bonds between Arg-620, Asp704 and their backbones.

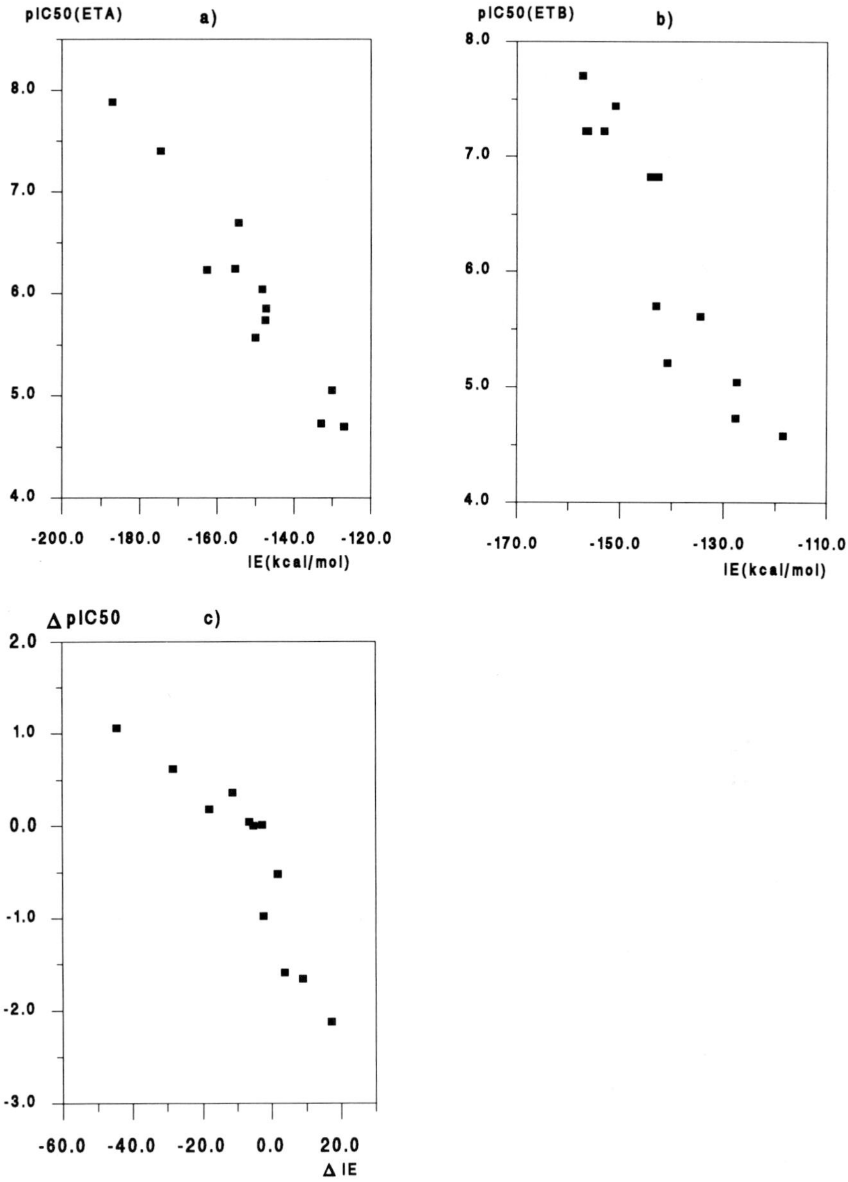

Figure 5. Correlation between the binding affinity data and the calculated interaction energies for the minimized peptide antagonist–receptor complexes. The linear regressions are: (a) $pIC_{50}(ETA) = -2.10(\pm0.70) - 0.054(\pm0.005)IE$, $n = 12$, $r = 0.965$, $s = 0.270$; (b) $pIC_{50}(ETB) =- 6.78(\pm1.60) - 0.09(\pm0.01)IE$, $n = 12$, $r = 0.933$, $s = 0.398$; (c) $\Delta pIC_{50} =- 0.77(\pm0.16) - 0.05(\pm0.008)\Delta IE$, $n = 12$, $r = 0.882$, $s = 0.491$.
where, n is the number of compounds, r is the correlation coefficient, s is the standard deviation and the values in parentheses give the 95% confidence intervals.

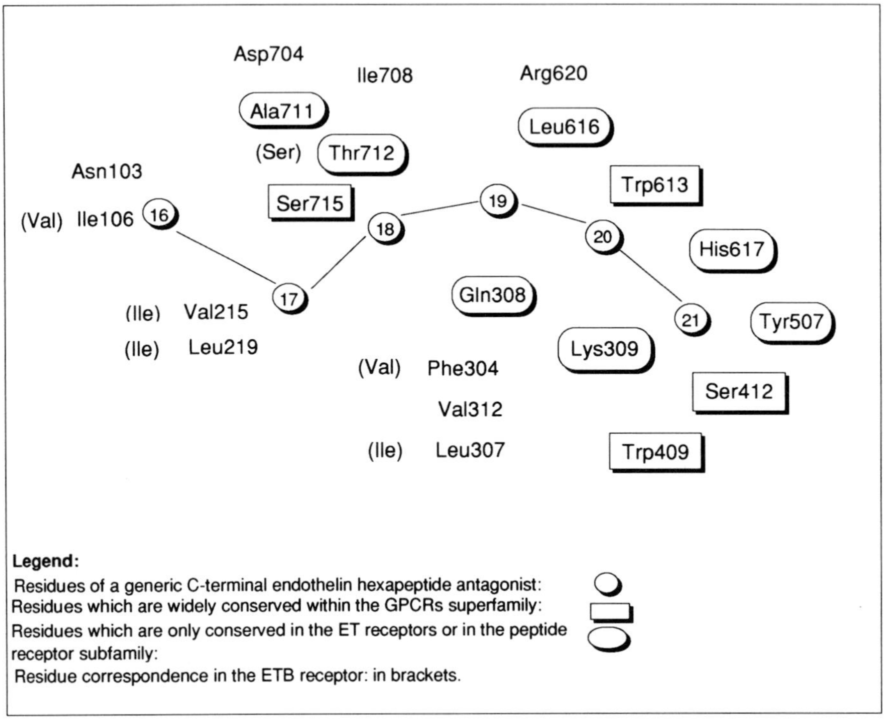

Figure 6. Scheme of the transmembrane amino acid residues of the endothelin receptor subtypes which are involved in the interaction with the twelve peptide antagonists studied.

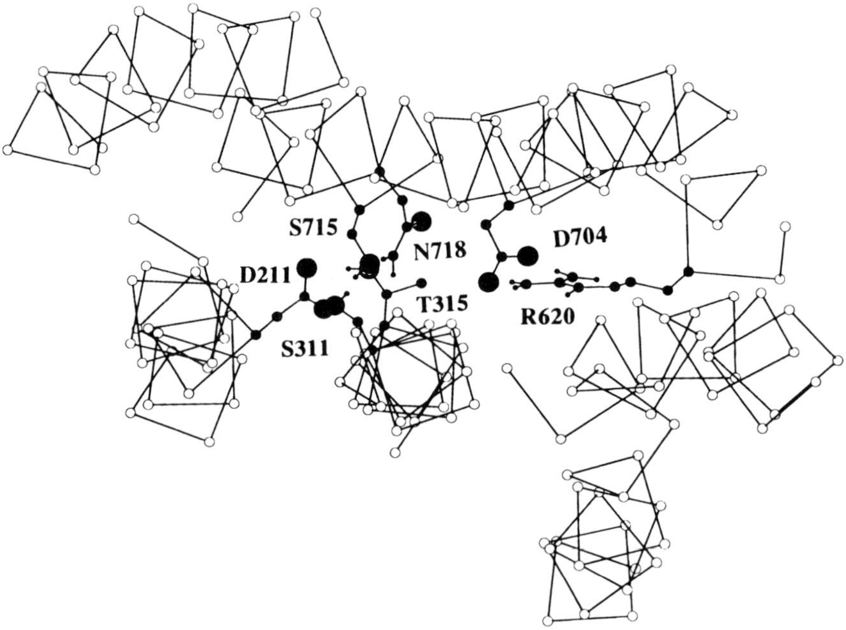

Figure 7. Inter-helix hydrogen bond network established in the isolated ETA receptor.

Table 2. Experimental binding affinities (pKi) and computed interaction energies (IE, kcal mol^{-1}) for the 1,3-diarylindan-2-carboxylic acid antagonists complexed with the ETA and ETB receptor subtypes

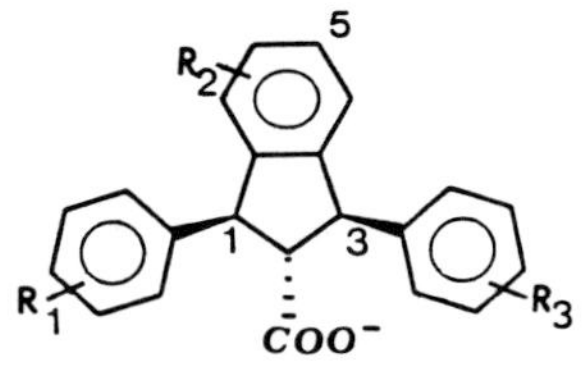

N	R-1	R-2	R-3	ETA pKi	ETA IE	ETB pKi	ETB IE
1	3,4-methylenedioxy	5-O-n-Pr	2-carboxymethoxy, 4-OMe	9.37	−118.58	7.82	−100.29
2	H	H	H	4.95	−69.61	4.52	−70.34
3	4-OMe	H	H	5.25	−75.48	4.52	−75.48
4	4-OMe	H	4-OMe	6.38	−84.03	5.61	−80.64
5	4-OMe	H	3,4-methylenedioxy	7.34	−86.44	6.12	−83.86
6	3,4-methylenedioxy	5-OH	4-OMe	7.82	−93.00	6.32	−84.90
7	3,4-methylenedioxy	5-O-n-Pr	4-OMe	7.95	−100.26	5.98	−84.34

Non-peptide ETA and ETB receptor antagonists

The results obtained for the C-terminal endothelin hexapeptide antagonists have provided us with a precious tool to guide the formulation of peptido-mimetic hypotheses for the docking of the 1,3-diarylindan-2-carboxylic acid series of potent ETA selective antagonists, reported by Elliott *et al.* (1994). In their original article, the authors are inclined towards a peptido-mimetic hypothesis which identifies residues Tyr-13, Phe-14, Asp-18 and the C-terminal carboxyl group of ET-1 as possible counterparts of the functional groups of compound 1 in *Table 2* (SB 209670).

On the basis of the knowledge gained by the analysis of the changes in conformation, topography and dynamics important for the interaction of the antagonist peptides with their ET receptors, we have proposed a different peptido-mimetic hypothesis based on the intermolecular interactions of three appropriately oriented binding groups and three key binding site residues, as shown in *Figure 8*.

Very good correlations have been obtained between the binding affinity data (pKi) (Elliott *et al.*, 1994) and the computed interaction energies (IE) (Menziani *et al.*, 1995b), see *Figure 9*(*a*) and (*b*). In addition, the selectivity of these compounds towards the ETA receptor subtype is successfully elucidated by the QSAR model reported (*Figure 9*(*c*)).

A comparison of the interaction preferences of the peptide antagonist PD142893 (compound 9, *Table 1*) and of the non-peptide antagonist SB 209670 (compound 1, *Table 2*) with the ETA binding site residues, resulting from QMD minimization

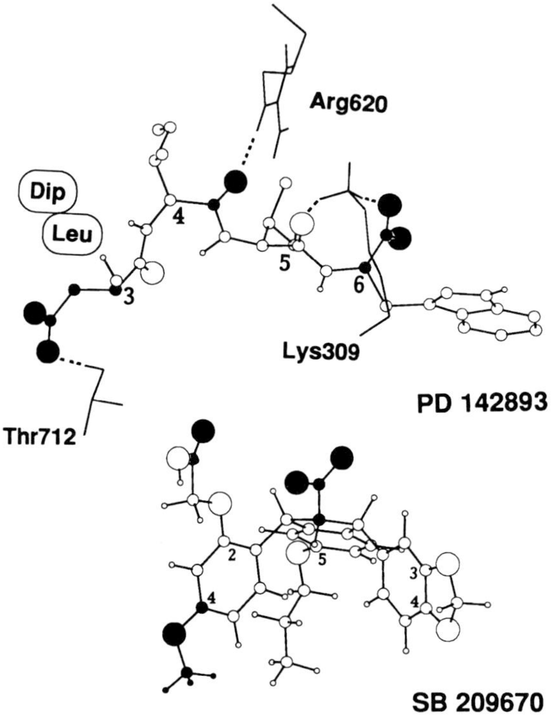

Figure 8. Peptido-mimetic hypothesis for the SB 209670 antagonist.

(*Figures 10* and *11*, see page 107) shows that, notwithstanding the fact that the transmembrane residues that participate in the binding of the two antagonists are basically the same, significant differences in their conformations are observed. However, quantitative analyses of the two complexes reveal great similarities in the interaction energies of these compounds with the ETA receptor subtype, when the portion of the peptide antagonist (residues 3 to 6) mimicked by SB 209670 is considered. The results of this comparison are reported in *Table 3*. The total interaction energy of the SB 209670 and of the PD 142893[3–6] antagonists are reported together with their van der Waals, electrostatic and hydrogen bond components. It is worth noticing that the multifaceted character of the amino acid residues of the PD 142893[3–6] portion is very well reproduced by the capability of SB 209670 to establish specific and dispersive interactions by means of the three functional groups highlighted in *Figure 8* (which can be described by the electrostatic and hydrogen bond contributions, in *Table 3*) and of the aromatic skeleton (which can be described by the van der Waals contribution, in *Table 3*), respectively. The contribution of each amino acid residue of the PD 14289[3–6] portion to the total interaction energy of this peptide antagonist is also listed. From the receptor

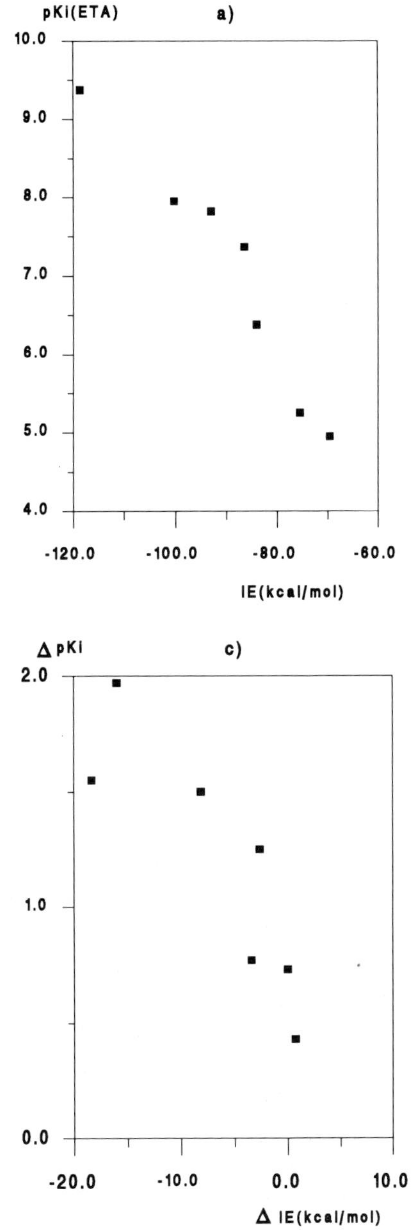

Figure 9. Correlation between the binding affinity data and the calculated interaction energies for the minimized non-peptide antagonist–receptor complexes. The linear regressions are: (a) pKi(ETA) = –1.34(±1.03) – 0.093(±0.011)IE, $n = 7$, $r = 0.965$, $s = 0.456$, $F = 67.08$; (b) pKi(ETB) = – 4.02(±0.91) – 0.12(±0.01)IE, $n = 7$, $r = 0.979$, $s = 0.252$, $F = 117.64$; (c) ΔpKi = 0.75(±0.16) 0.06(±0.01)ΔIE, $n = 7$, $r = 0.864$, $s = 0.301$, $F = 14.79$.

Table 3. Contributors to the interaction energies (IE, kcal mol^{-1}) of the complexes between the ETA receptor and the PD 142893[3-6] and SB 209670 antagonists

Antagonists	IE	E(vdW)	IE(El)	IE(HB)	Receptor residues	IE	IE(vdW)	IE(El)	IE(HB)
					Lys-309	−16.35	−4.23	−5.52	−6.59
SB 209670	−118.51	−70.58	−17.81	−30.12	Arg-620	−18.36	−3.39	−6.49	−6.59
					Thr-714	−5.25	−2.19	−0.09	−2.97
					Lys-309	−19.29	−4.69	−11.07	−3.51
PD 16289[3-6]	−119.68	−79.15	−21.94	−23.39	Arg-620	−12.52	−5.49	−3.60	−3.41
					Thr-712	−8.00	−2.00	−2.07	−3.92
Asp-3	−23.68	−14.34	−2.29	−7.04					
Ile-4	−24.81	−18.13	−2.92	−5.76					
Ile-5	−19.41	−16.10	−3.17	−0.13					
Trp-6	−51.78	−30.76	−13.56	−10.46					

point of view, our peptido-mimetic hypothesis is nicely supported by the similarity in the interaction energies of the three key binding residues initially individualized by means of a detailed molecular dynamics analysis of the binding preferences of the peptide antagonists (see *Table 3*). In particular, Lys-309 and Arg-620 achieve, in the interaction with SB 209670, an optimum orientation and establish very good hydrogen bonds with the carboxylic groups of the inhibitor. Thr-712 is substituted for Thr-714 in the complex with SB 209670. In fact, during the molecular dynamics simulation Thr-714 finds a more appropriate orientation to interact, through the establishment of a permanent hydrogen bond, with the para-methoxy group of SB 209670, than does Thr-712.

5. Conclusions

The three-dimensional chemical formalism applied to the receptor description in the context of QSAR analysis provides a quantitative and consistent tool for comparisons between the different receptor subtypes. In this way, receptor structure affinity-selectivity relationships can be comparatively elucidated. Moreover, modelling the conformational behaviour of the ligands interacting with their receptors by means of MD simulations and analysing their structure/dynamic/activity differences provide us with a framework for understanding and rationalizing the experimental data and guiding the design of new compounds with improved characteristics. Finally, the results obtained reveal the potential of the computational strategy adopted as a rational design approach to peptide mimetics.

However, verification by, or consistency with, an independent approach is required in order to extend the level of confidence ascribable to any result or prediction. In this respect, biophysical protocols designed on the basis of the QSARs obtained might provide good opportunities to support, improve, refine or modify the models proposed.

References

Adachi M, Furuichi Y, Miyamoto C. (1994) Identification of ligand-binding site of the human endothelin-A receptor and specific regions required for ligand selectivity. *Eur. J. Biochem.* **220**: 37–43.

Ballesteros JA, Weinstein H. (1994) Integrated methods for the construction of three dimensional models and computational probing of structure-function relations in G-protein coupled receptors. *Methods Neurosci.* **25**: 366–428.

Becker A, Haendler B, Hechler U, Schleuning WD. (1994) Mutational analysis of human endothelin receptors ETA and ETB identification of regions involved in the selectivity for endothelin-3 or cyclo-(D-Trp-D-Asp-Pro-D-Val-Leu). *Eur. J. Biochem.* **221**: 951–958.

Bousso D, Bdolah A, Sokolowsky M. (1992) Involvement of tyrosyl residue(s) in binding of endothelin and sarafotoxin to their receptors in rat brain and heart. *Neurosci. Lett.* **140**: 247–250.

De Benedetti PG, Menziani MC, Fanelli F, Cocchi M. (1993) The heuristic-direct approach to quantitative structure–activity relationship analysis. *J. Mol. Struct. (Theochem)*, **285**: 147–153.

Doherty AM, Cody WL, DePue PL, He JX, Waite LA, Leonard DM, Leitz NL, Dudley DT, Rapundalo ST, Hingorani GP, Haleen SH, LaDouceur DM, Hill KE, Flynn MA, Reynolds EE. (1993) Structure–activity relationships of C-terminal endothelin hexapeptide antagonists. *J. Med. Chem.* **36**: 2585–2594.

Elliott JD, Lago MA, Cousins RD, Gao A, Leber JD, Erhard KF, Nambi P, Elshourbagy NA, Kumar C, Lee JA, Bean JW, DeBrosse CW, Eggleston DS, Brooks DP, Feuerstein G, Ruffolo RR Jr, Weinstock J, Gleason JG, Peishoff CE, Ohlstein EH. (1994) 1,2-diarylindan-2-carboxylic acids, potent and selective non-peptide endothelin receptor antagonists. *J. Med. Chem.* **37**: 1553–1557.

Fanelli F, Menziani MC, Carotti A, De Benedetti PG. (1994a) Theoretical quantitative structure-activity analysis on three dimensional models of ligand-m1 muscarinic receptor complexes. *Bioorg. Med. Chem.* **2**: 195–211.

Fanelli F, Menziani MC, Cocchi M, Leonardi A, De Benedetti PG. (1994b) The heuristic-direct approach to theoretical quantitative structure activity relationship analysis of α_1-adrenoceptor ligands. *J. Mol. Struct. (Theochem)*, **314**: 265–276.

Fanelli F, Menziani MC, Cocchi M, De Benedetti PG. (1995) Comparative molecular dynamics study on the seven-helix bundle arrangement of G-protein coupled receptors. *J. Mol. Struct. (Theochem)*, **333**: 46–69.

Fong TM, Yu H, Cascieri MA, Underwood D, Swain CJ, Strader CD. (1994) The role of histidine 265 in antagonist binding to the neurokinin-1 receptor. *J. Biol. Chem.* **269**: 2728–2732.

Greer J, Erickson JW, Baldwin JJ, Varney MD. (1994) Application of the three-dimensional structures of protein target molecules in structure-based drug design. *J. Med. Chem.* **37**: 1035–1054.

Hollenberg MD. (1990) Receptor-membrane interactions: a new focus for structure-activity relationships. In: *Trends in Drug Research* (ed. V. Claassen). Amsterdam: Elsevier, pp. 11–22.

Huggins JP, Pelton JT, Miller RC. (1993a) The structure and specificity of endothelin receptors: their importance in physiology and medicine. *Pharmacol. Ther.* **59**: 55–123.

Huggins JP, Trumpp-Kallmeyer S, Hibert MF, Hoflack JM, Fanger BO, Jones CR. (1993b) Modelling and modification of the binding site of endothelin and other receptors. *Eur. J. Pharmacol.* **245**: 203–214.

Krystek SR, Patel PS, Rose PM, Fisher SM, Kienzle BK, Lach DA, Liu ECK, Lynch JS,

Novotny J, Webb ML. (1994) Mutation of peptide binding site in transmembrane region of a G-protein coupled receptor accounts for endothelin receptor subtype selectivity. *J. Biol. Chem.* **269**: 12383–12386.

Lee JA, Elliott JD, Sutiphong JA, Friesen WJ, Ohlstein EH, Stadel JM, Gleason JG, Peishoff CE. (1994) Tyr-129 is important to the peptide ligand affinity and selectivity of human endothelin type A receptor. *Proc. Natl Acad. Sci. USA,* **91**: 7164–7168.

Masaki T, Vane JR, Vanhoutte PM. (1994) International union of pharmacology nomenclature of endothelin receptors. *Pharmacol. Rev.* **46**: 137–142.

Menziani MC, De Benedetti PG, Gago F, Richards WG. (1989) The binding of benzensulphonamides to carbonic anhydrase enzyme. A molecular mechanics study and quantitative structure-activity relationships. *J. Med. Chem.* **32**: 951–956.

Menziani MC, De Benedetti PG. (1991a) Direct and indirect theoretical QSAR modelling in sulphonamide carbonic anhydrase inhibitors. In: *QSAR: Rational Approaches on the Design of Bioactive Compounds* (Silipo C, Vittoria A, eds). Amsterdam: Elsevier, pp. 331–334.

Menziani MC, De Benedetti PG. (1991b) Molecular mechanics and quantum chemical QSAR analysis in carbonic anhydrase-heterocyclic sulphonamides interactions. *Struct. Chem.* **3**: 215–219.

Menziani MC, Cocchi M, De Benedetti PG, Gilbert RG, Richards WG, Zamai M, Caiolfa VR. (1991c) A theoretical study of the structure of big-endothelin. *J. Chem. Phys.* **88**: 2687–2694.

Menziani MC, Cocchi M, De Benedetti PG. (1993) A molecular dynamics simulation of sequence-directed recognition peptides interacting with big-Endothelin. *J. Mol. Struct. (Theochem),* **286**: 95–108.

Menziani MC, Cocchi M, Fanelli F, De Benedetti PG. (1995a) Quantitative structure-affinity/selectivity relationship analysis on three dimensional models of the complexes between the ETA and ETB receptors and C-terminal endothelin hexapeptide antagonists. *J. Mol. Struct. (Theochem),* **333**: 243–248.

Menziani MC, Cocchi M, Fanelli F, De Benedetti PG. (1994b) Theoretical QSAR analysis on three dimensional models of the complexes between peptide and non-peptide antagonists of the ETA and ETB receptors. In: *Concepts, Tools and Biological Applications* (Sanz F, ed.). S.A.: J.R. Prous.

Richards WG. (1983) *Quantum Pharmacology.* London: Butterworth.

Sakamoto A, Yanagisawa M, Sawamura T, Enoki T, Ohtani T, Sakurai T, Nakao K, Toyo-oka T, Masaki T. (1993) Distinct subdomains of human endothelin receptors determine their selectivity to endothelin-A-selective antagonist and endothelin-B-selective agonists. *J. Biol. Chem.* **268**: 8547–8553.

Savarese TM, Fraser CM. (1992) In vitro mutagenesis and the search for structure-function relationships among G-protein coupled receptors. *Biochem. J.* **283**: 101–119.

Sjostrom M, Wold S. (1981) Linear free energy relationships. Local empirical rules or fundamental laws of chemistry? *Acta Chem. Scand.* **B 35**: 537–554.

Trumpp-Kallmeyer S, Hoflack JM, Bruinvels A, Hibert MF. (1992) Modelling of G-protein coupled receptors: application to dopamine, adrenaline, serotonin, acetylcholine and mammalian opsin receptors. *J. Med. Chem.* **35**: 3448–3462.

Zhang D, Weinstein H. (1993a) Ligand selectivity and the molecular properties of the 5-HT_2 receptor: computational simulations reveal a major role for transmembrane helix 7. *Med. Chem. Res.* **3**: 357–369.

Zhang D, Weinstein H. (1993b) Signal trasduction by a 5-HT2 receptor: a mechanistic hypothesis from molecular dynamics simulations of the three-dimensional model of the receptor complexed to ligands. *J. Med Chem.* **36**: 934–938.

Zhu G, Wu LH, Mauzy C, Egloff AM, Mirzadegan T, Chung FZ. (1992) Replacement of Lys-181 by aspartic acid in the third transmembrane region of endothelin type B receptor reduces its affinity to endothelin peptides and sarafotoxin S6C without affecting G-protein coupling. *J. Cell Biochem.* **50**: 159–164.

Zoffmann S, Gether U, Schwartz TW. (1993) Conserved His (VI-17) of the NK-1 receptor is involved in binding of non-peptide antagonists but not substance P. *FEBS Lett.* **336**: 506–510.

Chapter 8

Molecular modelling of β-adrenoceptors

A.P. IJzerman and Helene M. Zuurmond

1. Introduction

The β-adrenoceptor is the protein target in the human body for such diverse drugs as isoprenaline (isoproterenol), salbutamol, terbutaline (all three agonists used in the threatment of asthma), propranolol and pindolol (both antagonists used in the treatment of hypertension) (*Figure 1*). The endogenous ligands for the β-adrenoceptors are the hormone adrenaline (epinephrine) and the neurotransmitter noradrenaline (norepinephrine). Like the synthetic agonists, they generate a signal that is transduced via G proteins into the cell leading to the enhanced production of cyclic AMP by the enzyme adenylate cyclase. This cascade of events has been largely identified with one of the subtypes, the β_2-adrenoceptor, as the prototypical receptor. It appears to be a general mechanism for receptor activation (for reviews on the β-adrenoceptor see Kobilka, 1992 and Ostrowski *et al.*, 1992).

Identification of the ligand binding site is an important aspect of the drug–receptor interaction. With no experimentally determined receptor structure available, circumstantial evidence becomes the sole source for facts and fiction. Two complementary approaches can be distinguished. First, information embedded in the ligands (agonists and antagonists) can be used to obtain a fingerprint of their binding site on the receptor. Biological data, such as affinity values obtained from radioligand binding studies on wild-type and mutated receptors, and (physico)-chemical parameters defining the spatial, electronic and lipophilic characteristics of the compounds, are both of invaluable help. A second, more direct, approach is the construction of receptor models based on the putative resemblance of G protein-coupled receptors (GPCRs) and bacteriorhodopsin, of which experimentally determined atomic coordinates are available (Henderson *et al.*, 1990).

It is the thrust of this chapter that the two approaches should be combined and used simultaneously to come to verifiable hypotheses. Quantitative structure–affinity relationships (QSARs) will be derived for both agonists and antagonists on the β-adrenoceptor. Based on the emerging information, receptor models will be generated.

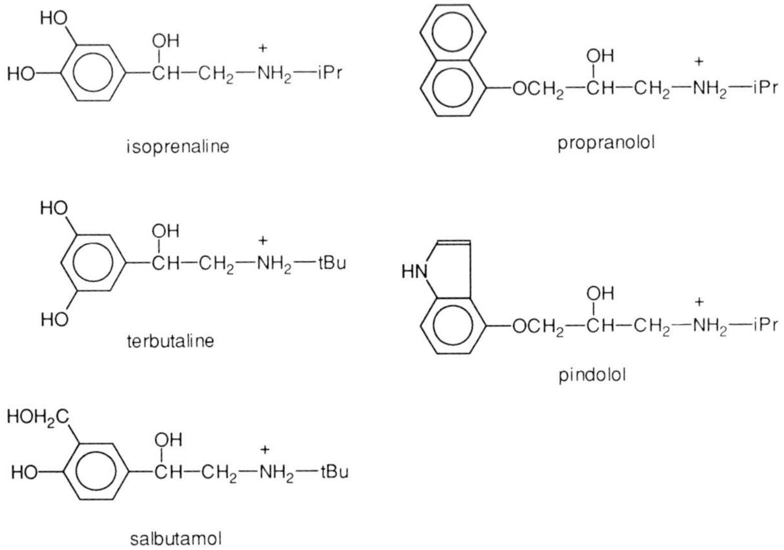

Figure 1. Chemical structures of β-adrenoceptor ligands.

2. Methods

QSAR parameters and analysis

Log *P* values (octanol/water) of the aromatic moiety of the ligands (i.e. the corresponding, substituted benzenes) were calculated (Rekker and De Kort, 1979). The steric branching parameter (Sb) for all aromatic substituents was calculated according to Austel *et al.* (1979). Computer-assisted multiple regression analyses were performed, which yielded the regression equations together with statistical parameters, adjusted for the degrees of freedom. Regression coefficients are given with their 95% confidence intervals. Further details have been described by IJzerman *et al.* (1985).

β_2-adrenoceptor affinity

K_i values (as a measure for β_2-adrenoceptor affinity) were determined by computer-assisted analysis of the inhibition of the specific (–)-[^{3}H]dihydroalprenolol (DHA) binding at pH 7.5 to the β_2-adrenoceptors of a bovine skeletal muscle membrane preparation (IJzerman *et al.*, 1984b)

Receptor modelling

The method described by Lewell (1992) was essentially followed to generate a model of the transmembrane domains of the human β_2-adrenoceptor. Model building, ligand docking, and energy minimizations were carried out using the software package BIOGRAF, version 3.1 (Molecular Simulations Inc., Waltham, MA, USA). All manipulations were performed on a Silicon Graphics 4D/25GT or 4000XZ work-station. The atomic coordinates of bacteriorhodopsin were retrieved from the Brookhaven Protein Databank (reference code 1BRD). The sequence alignment of the transmembrane segments of bacteriorhodopsin with those of bovine opsin and the human β_2-adrenoceptor was identical to the one presented by Lewell (1992). All problems and caveats associated with the modelling procedure have been extensively documented previously in our work on the modelling of adenosine receptors (IJzerman *et al.*, 1992; 1994).

3. Results and discussion

QSAR studies

The K_i values of two classes of β-adrenoceptor agonists (chemically both phenylethanolamines) and one class of β-adrenoceptor antagonists (chemically phenoxypropanolamines) are listed in *Tables 1–3*, respectively. Since the cationic species, with a protonated nitrogen atom in the aliphatic side chain, govern β_2-adrenoceptor affinity (IJzerman *et al.* 1984a,b), all experimentally determined K_i values have been adjusted for the amounts of cations present at pH 7.5, which is especially important in the class of the phenylethanolamines, usually dibasic acids. Moreover, in case of racemates, pK_i values have been raised with 0.3 (log 2), as only (–)-isomers are considered to contribute to β-adrenoceptor affinity. As an extreme example, VUF 8303 (*Table 2*) has a K_i value of only 1.95 μM, which nevertheless generated a high 'corrected' pK_i value of 7.41, since the compound is a racemate and has only 4% cations at pH 7.5.

If the influence of the aliphatic side chain on β-adrenoceptor affinity is assumed to be identical within one class, it should be possible to correlate physicochemical properties of the aromatic moieties only with the affinities of the whole molecules. Thus, by considering the β-adrenoceptor ligands as the corresponding substituted benzenes, a large and diverse parameter set was easily obtained. This would have been far more difficult or even impossible with respect to the 'intact' derivatives, never mind the experimental problems associated with the determination of lipophilicities of charged molecules (i.e. cations). We calculated the log P values (as a measure for hydrophobicity) of all substituted benzenes (*Table 4*) and found them to correlate well with the known experimentally determined ones (data not shown). As a steric parameter we selected the steric branching parameter (*Sb*, see *Table 4*) developed by Austel *et al.* (1979). Although *Sb* is a somewhat intuitively developed parameter, its application is favoured by its easy calculation for any conceivable substituent.

Table 1. β_2-adrenoceptor affinities of substituted *N*-isopropylphenylethanolamines (iPr: isopropyl)

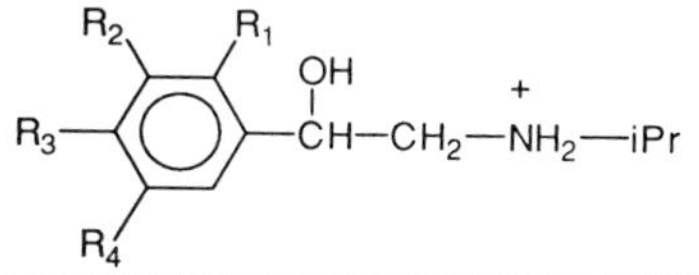

Compound	R_1	R_2	R_3	R_4	Apparent K_i (μM)	Corrected pK_i
Isoprenaline	H	OH	OH	H	0.33	6.51
Orciprenaline	H	OH	H	OH	9.12	5.37
Du 21117	H	NH_2	OH	H	5.13	5.59
AH 3021	H	CH_2OH	OH	H	3.16	5.80
NAB 277	H	Cl	NH_2	Cl	0.058	7.54
N-iPr-norsynephrine	H	H	OH	H	4.27	5.67
Pronethalol	H	–CH=CH–CH=CH–		H	0.16	7.10
INPEA	H	H	NO_2	H	1.32	6.18
Sotalol	H	H	$NHSO_2CH_3$	H	0.58	6.57
Ni-iPr-norphenylephrine	H	OH	H	H	0.91	6.34

Table 2. β_2-adrenoceptor affinities of substituted *N*-*t*-butylphenylethanolamines (tBu: tert.butyl)

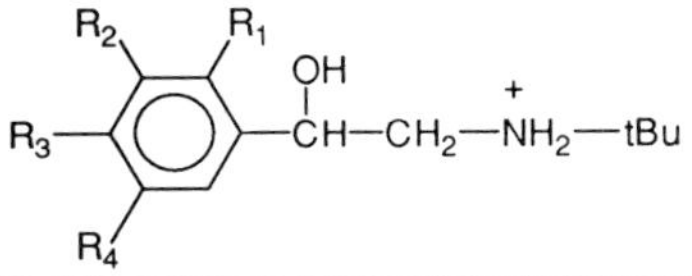

Compound	R_1	R_2	R_3	R_4	Apparent K_i (μM)	Corrected pK_i
Th 1206	H	OH	OH	H	0.60	6.55
Terbutaline	H	OH	H	OH	3.98	5.73
Du 28663	H	NH_2	OH	H	2.04	5.99
SKF 56301	H	$NHCH_3$	OH	H	0.52	6.58
Salbutamol	H	CH_2OH	OH	H	0.52	6.58
Clenbuterol	H	Cl	NH_2	Cl	0.036	7.74
N-tBu-norsynephrine	H	H	OH	H	1.26	6.20
AH 3474	H	$CONH_2$	OH	H	0.47	7.03
C 78	Cl	H	H	H	0.13	7.18
VUF 8303	H	Cl	OH	Cl	1.95	7.41

Lipophilicity of the aromatic ring system in the compounds proved to be important for affinity. The regression equation for the *N*-*t*-butylphenylethanolamines was

$$pK_i = 0.53(\pm 0.12) \log P + 5.87(\pm 0.23)$$

$$n = 10,\ r = 0.84,\ s = 0.37,\ F = 18.5$$

Table 3. β_2-adrenoceptor affinities of substituted *N*-isopropylphenoxypropanolamines (iPr: isopropyl)

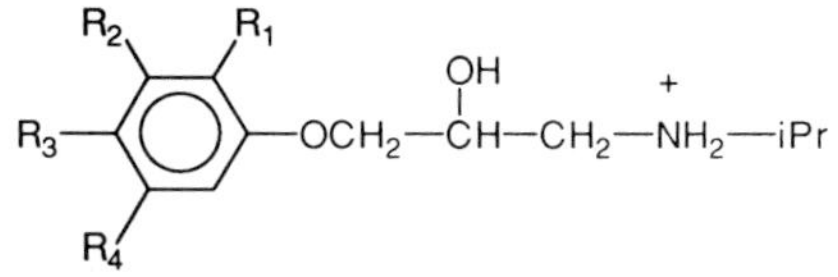

Compound	R_1	R_2	R_3	R_4	Apparent K_i (μM)	Corrected pK_i
Propranolol	–CH=CH–CH=CH–		H	H	0.89	9.05
Alprenolol	CH_2CHCH_2	H	H	H	0.39	9.41
Dihydroalprenolol	$CH_2CH_2CH_3$	H	H	H	0.63	9.20
Practolol	H	H	$NHCOCH_3$	H	20,417	4.99
Kö 589	CH_3	H	H	H	10	8.29
Kö707	H	CH_3	H	CH_3	10	8.29
Kö 1124	H	$CH_3CHCH_2H_5$	H	H	18	8.05
Kö 1313	CN	H	H	H	22	7.96
Kö 1350	CH_2OH	H	H	H	50	7.60
Kö 592	H	CH_3	H	H	44	7.66
—	H	H	H	H	54	7.57
Kö 1411	OCH_2CCH	H	H	H	2.1	8.97
Prenalterol	H	H	OH	H	1,072	6.27
Betaxolol	H	H	$C_2H_4OCH_2cPr$	H	186	7.03

In this particular case the contribution of lipophilicity to binding explains more than 70% of the observed variance (r^2 = 0.71). All equations, however, could still be improved by the introduction of the steric parameter. Other parameters such as the electronic Hammett's σ were tried too, but proved to be unsuccessful. As the regression coefficients for the substituents R_2 and R_3 (*meta* and *para* to the aliphatic side chain, respectively) appeared to be approximately identical within one class, the respective parameter values were summed and used in the following equations for all three classes of compounds.

N-*t*-butylphenylethanolamines (agonists):

$$pK_i = 0.63(\pm 0.07) \log P + 0.27(\pm 0.06)\, Sb_{R2+R3} + 5.17(\pm 0.20)$$

$$n = 10,\ r = 0.95,\ s = 0.22,\ F = 40.6$$

N-isopropylphenylethanolamines (agonists):

$$pK_i = 0.61(\pm 0.13) \log P + 0.20(\pm 0.10)\, Sb_{R2+R3} + 4.91(\pm 0.32)$$

$$n = 10,\ r = 0.87,\ s = 0.39,\ F = 12.6$$

N-isopropylphenoxypropanolamines (antagonists):

$$pK_i = 0.94(\pm 0.13) \log P + 0.32(\pm 0.11)\, Sb_{R1} - 0.28(\pm 0.07)\, Sb_{R2+R3} + 5.52(\pm 0.34)$$

$$n = 14,\ r = 0.95,\ s = 0.44,\ F = 35.3$$

Table 4. Physicochemical parameters of substituted benzenes

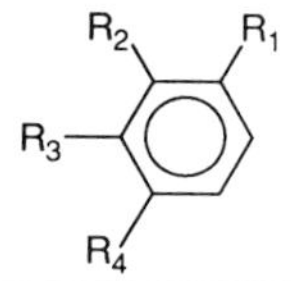

R_1	R_2	R_3	R_4	log P_{calc}	Sb_{R1}	Sb_{R2+R3}
H	H	H	H	2.02	0.0	0.0
H	OH	OH	H	1.03	0.0	2.0
H	OH	H	OH	0.74	0.0	1.0
H	OH	H	H	1.53	0.0	1.0
H	H	OH	H	1.53	0.0	1.0
H	NH_2	OH	H	0.50	0.0	2.0
H	$NHCH_3$	OH	H	1.10	0.0	3.0
H	CH_2OH	OH	H	0.68	0.0	3.0
H	$CONH_2$	OH	H	1.10	0.0	4.0
H	Cl	OH	Cl	3.01	0.0	2.2
Cl	H	H	H	2.76	1.2	0.0
H	Cl	NH_2	Cl	3.06	0.0	2.2
–CH=CH–CH=CH–		H	H	3.30	1.0	1.0
H	–CH=CH–CH=CH–	H	3.30	0.0	2.0	
H	H	NO_2	H	1.79	0.0	3.0
H	H	$NH_2SO_2CH_3$	H	0.95	0.0	5.2
CH_2CHCH_2	H	H	H	3.22	3.0	0.0
$CH_2CH_2CH_3$	H	H	H	3.58	3.0	0.0
H	H	$NHCOCH_3$	H	1.25	0.0	4.0
CH_3	H	H	H	2.54	1.0	0.0
H	CH_3	H	CH_3	3.06	0.0	1.0
H	$CH_3CHC_2H_5$	H	H	4.10	0.0	4.0
CN	H	H	H	1.67	2.0	0.0
CH_2OH	H	H	H	1.18	2.0	0.0
H	CH_3	H	H	2.54	0.0	1.0
OCH_2CCH	H	H	H	2.41	3.0	0.0
H	H	$C_2H_4OCH_2cPr$	H	3.18	0.0	3.0

What can be learnt from these equations?

First of all, lipophilicity (log *P*) is an important parameter in β_2-adrenoceptor affinity. From the equations it follows that the more lipophilic the aromatic nucleus of the compounds is, the higher the affinity will be. The extent, however, is different between agonists and antagonists. The two classes of the phenylethanolamines (all agonists) have identical regression coefficients for log *P* (0.63 and 0.61, respectively), which are significantly lower than the one for the antagonistic phenoxypropanolamines (0.94). In terms of receptor fingerprints this could mean that the aromatic moieties of agonists are in an environment that is different from, and less lipophilic than, the region where the aromatic ring system of antagonists binds. This hypothesis is strongly supported by the relative importance of the substituents on the aromatic rings. Clearly, the steric requirements for the phenoxypropanolamines are in contrast with those for both classes of phenylethanolamines. Substitution of H by R_2 or R_3 is unfavourable in the antagonists (the regression coefficient is –(!)0.28), whereas this substitution pattern is advantageous in the agonists (regression coefficient is +0.27 or +0.20). *Ortho* (R_1) substitution does not seem to be sterically limited in the phenoxypropanolamines. It enhances β_2-adrenoceptor affinity due to the introduction of both extra lipophilicity and steric 'bulk' in the compounds. Since only one *ortho* substituent is found in the phenylethanolamines, the importance of this position is unclear. The results described so far are qualitatively summarized in *Figure 2.*

These results question all ligand models in which the aromatic nuclei of agonists and antagonists coincide. The apparent assumption in such models is that for a competitive interaction a maximal degree of overlap between agonists and antagonists is essential. Thus, Jen and Kaiser (1977) have proposed a 'rigid' bicyclic structure for the phenoxypropanolamines, which can be superimposed on the phenylethanolamines in a way that both phenyl rings are in the same position. More recently, Strosberg *et al.* (1993) superimposed the aromatic moieties as well as the protonated amino function of both agonists and antagonists when docking these compounds in their β_2-adrenoceptor model.

The intercepts in the above equations are measures of the importance of the side chains for receptor affinity. Since log *P* and *Sb* were parameters used to describe the aromatic nuclei of the compounds only, the contribution of the side chain with the protonated amino function and the β-OH group (at the chiral carbon atom) is reflected in the intercepts. The 'extra' oxymethylene bridge in the oxypropanolamine compared with the ethanolamine side chain may be responsible for the higher value of 5.52, compared with 4.91 and 5.17 for the two classes of phenylethanolamines. Furthermore, this ether linkage may direct the aromatic ring system of the antagonists into a more hydrophobic region, as described above. This raises the issue whether the hydroxy group and the protonated nitrogen, present in all compounds, interact with receptor residues that are the same for agonists and antagonists.

Receptor modelling

The amino acid sequence of the human β_2-adrenoceptor together with its putative membrane topology is depicted in *Figure 3*. Based on the atomic coordinates of bacteriorhodopsin a homologous three-dimensional model of the membrane α-helices of the receptor was built. The overall architecture of the receptor model was virtually identical to that of bacteriorhodopsin because of the procedure followed

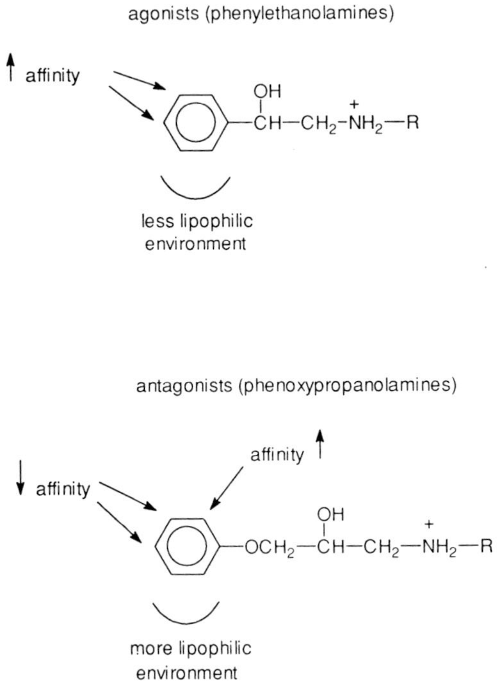

Figure 2. Qualitative summary of QSAR analysis of β-adrenoceptor agonists (phenylethanolamines) and antagonists (phenoxypropanolamines).

(i.e. 'on screen' mutation and subsequent minimization). It should be mentioned here that five proline residues, generally considered as 'helix beakers', are present in the transmembrane domains of both bacteriorhodopsin and the β_2-adrenoceptor, but not all at the same location. Thus, the receptor model can only be a crude approximation of biochemical reality, and it is best regarded as a visualization of current knowledge.

This is particularly true for the docking of ligands into the pore formed by the seven helices. The binding site for the highly hydrophobic ligand (and 'agonist') retinal in bacteriorhodopsin has been shown to be buried in the membrane, just above the centre of the cavity formed by the seven transmembrane domains. Are the hydrophilic β-adrenoceptor agonists in a similar position? From site-directed mutagenesis studies on the β_2-adrenoceptor some salient features have emerged that appear to confirm this hypothesis. First of all, substitution of Asp-113 (roughly halfway along helix III) with asparagine led to dramatic decreases in both agonist and antagonist affinity. Since other GPCRs that bind biogenic amines such as dopamine, acetylcholine etc., also have this aspartate present, it was concluded that this residue was the counter-ion for the protonated amino function in the ligands

(Strader *et al.*, 1987). Similarly, two serine residues in helix V (204 and 207) were implicated in the coordination of the catechol moiety in the agonist isoprenaline via two hydrogen bonds without significant effects on antagonist binding (Strader *et al.*, 1989). Asn-312 (helix VII) has also been reported to be important for the binding of the antagonistic phenoxypropanolamines (Suryanarayana and Kobilka, 1993). It was proposed that this asparagine residue forms a hydrogen bond with the oxygen atom in the oxymethylene bridge. We used all this information to dock isoprenaline and pindolol in the receptor cavity, as representative examples of β-adrenoceptor agonists and antagonists, respectively. Thus, it appeared relatively straightforward to position the isoprenaline molecule in such a way that the two phenolic hydroxy groups of the catechol function interact with the two serine residues on helix V. A hydrogen bond is formed between the oxygen atom of Ser-204 and the hydrogen atom of the *meta*-hydroxy group of isoprenaline. A similar hydrogen bond occurs

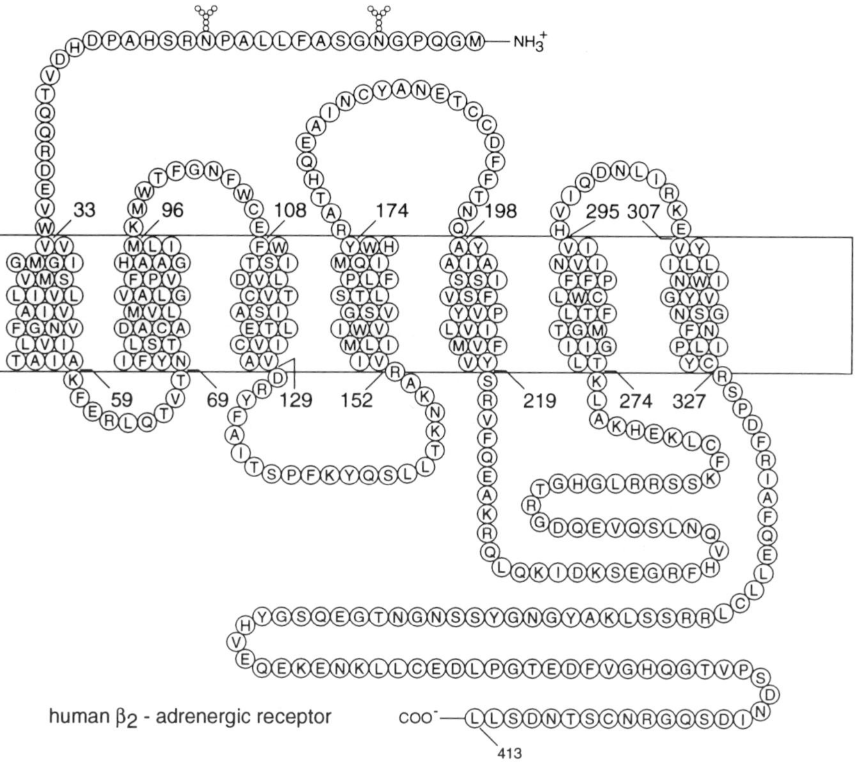

Figure 3. Two-dimensional representation of the human β_2-adrenoceptor. The N-terminus (with two glycosylation sites) is located on the extracellular side, and the C-terminus on the cytosolic side. Amino acids are indicated by the one-letter-notation.

between Ser-207 and isoprenaline's *para*-hydroxy group. An ionic interaction is seen between the protonated amino group of isoprenaline and the negatively charged aspartate residue 113 in helix III. This is summarized in *Figure 4a* (overview) and *Figure 4b* (detail) (see page 108). A similar procedure can be applied to the antagonist pindolol. Again, it appeared feasible to visualize the mutation data. Pindolol by virtue of its positively charged amino group also - interacts with Asp-113 , whereas Asn-312 may form some kind of double hydrogen bond with the β-hydroxy group and the oxygen atom of the ether function (*Figure 5a* and *5b* (see pages 108 and 109 respectively)). From this it follows that according to the modelling results agonists (phenylethanolamines) and antagonists (phenoxy-propanolamines) only partially overlap, as shown in Figure 6 (see page 109). All the data presented so far are in agreement with the modelling study by Lewell (1992).

When we consider in more detail all amino acids that surround the two molecules within 4 Å (*Figures 7* and *8*; see pages 109 and 110, respectively), it is readily apparent that the indol nucleus of pindolol is in a rather hydrophobic region, consisting of tryptophan, methionine, phenylalanine and tyrosine residues. This is unlike the area where the catechol function of isoprenaline binds; the two serine residues (204 and 207) cause a more hydrophilic environment. These findings are in agreement with the QSAR data described above; the different regression coefficients for the contribution of log P to affinity can be explained as due to different receptor environments.

From the modelling studies a clue to understand the ligands' stereoselectivity also emerges. It is well known that the (–)-isomers of both agonists (R designation) and antagonists (S designation) have considerably higher affinity for β-adrenoceptors than the corresponding (+)-isomers. Depending on optical purity and receptor recognition stereoselectivity ratios can be as high as 1000 (Morris and Kaumann, 1984). This critical property, however, has not yet been attributed to specific receptor residues. A direct consequence of the proposed limited overlap between agonists and antagonists on the β-adrenoceptor is the unequivalence of the two β-hydroxy groups that determine the stereoselectivity. In our model Asn-312 (helix VII) interacts with the hydroxy group of pindolol, but not with the same functionality in isoprenaline. This may agree with site directed mutagenesis data on not only β-adrenoceptors but also $5HT_{1A}$ receptors that recognize phenoxy-propanolamines (Suryanarayana and Kobilka, 1993; Guan *et al.*, 1992).

Lewell (1992) proposed that a serine residue in helix IV (Ser-161) forms a hydrogen bond with the β-hydroxy group in agonists. However, mutation of this residue into alanine had no effect on agonist affinity (Strader *et al.*, 1989). Hibert *et al.* (1991) in their seminal modelling paper proposed Ser-165 instead, but here mutagenesis studies did not yield more definitive proof, since substitution into alanine failed to produce immunoreactive protein (Strader *et al.*, 1989). The conformational flexibility of isoprenaline while anchored via the catechol and amine functions was still considerable in our modelling efforts. Indeed, residues in helix IV, such as the two serines (161 and 165), can be brought near the ethanolamine side chain. On the other hand, residues in helix VI are also in close proximity. In particular an asparagine in position 293 was capable of forming a hydrogen bond with isoprenaline's β-hydroxy group (see *Figure 8*). It should be pointed out that this asparagine residue is not unique for β-adrenoceptors. With respect to the biogenic amine receptors it is consistently present in all three β-adrenoceptor subtypes, but also in the dopamine D_1 and D_5 , and the 5-$HT_{1C/2}$ receptors. Both dopamine and serotonin, however, lack a chiral centre in their aliphatic side chains. Nevertheless, on the basis of these modelling considerations, a collaborative project was started with M. Lohse (Gene Center, University of Munich, Germany) to mutate this residue and analyse the effects of this change. Preliminary results of the Asn-293 to Leu mutation show that the affinity of (–)-isoprenaline was significantly reduced, while the (+)-isomer retained its 'wild-type' affinity.

4. Conclusion

The combined use of classic and newer tools in computational chemistry (QSAR analysis and molecular modelling) allowed the definition of putative ligand binding sites on the β_2-adrenoceptor, based on physicochemical and biological data. The model suggests that agonists and antagonists for the β_2-adrenoceptor overlap only partially, still leading to a mutual displacement and to their well-established competitive interaction. Residues that may be responsible for the stereospecific recognition of the ligands have been identified. It is hoped that combined efforts in computational chemistry and molecular biology may lead to validation and optimization of the proposed model, eventually enabling the rational design of new chemical entities.

References

Austel V, Kutter E, Kalbfleisch W. (1979) A new easily accessible steric parameter for structure-activity relationships. *Arzneim.-Forsch./Drug Res.* **29**: 585–587.

Guan XM, Peroutka SJ, Kobilka BK. (1992) Identification of a single amino acid residue for the binding of a class of β-adrenergic receptor antagonists to 5-hydroxytryptamine receptors. *Mol. Pharmacol.* **41**: 695–698.

Henderson R, Baldwin JM, Ceska TA, Zemlin F, Beckmann E, Downing KH. (1990) Model for the structure of bacteriorhodopsin based on high-resolution electron cryo-microscopy. *J. Mol. Biol.* **213**: 899–929.

Hibert MF, Trumpp-Kallmeyer S, Bruinvels A, Hoflack J. (1991) Three-dimensional models of neurotransmitter G-binding protein-coupled receptors. *Mol. Pharmacol.* **40**: 8–15.

IJzerman AP, Bultsma T, Timmerman H, Zaagsma J. (1984a) The ionization of β-adrenoceptor agonists: a method for unravelling ionization schemes. *J. Pharm. Pharmacol.* **36**: 11–15.

IJzerman AP, Bultsma T, Timmerman H, Zaagsma J. (1984b) The relation between ionization and affinity of β-adrenoceptor ligands. *Naunyn-Schmiedeberg's Arch. Pharmacol.* **327**: 293–298.

IJzerman AP, Aué) GHJ, Bultsma T, Linschoten MR, Timmerman H. (1985) Quantitative evaluation of the β_2-adrenoceptor affinity of phenoxypropanolamines and phenylethanolamines. *J. Med. Chem.* **28**: 1328–1334.

IJzerman AP, Van Galen PJM, Jacobson KA. (1992) Molecular modeling of adenosine receptors. I. The ligand binding site on the A_1 receptor. *Drug Design Discovery,* **9**: 49–67.

IJzerman AP, Van der Wenden EM, Van Galen PJM, Jacobson KA. (1994) Molecular modeling of adenosine receptors. The ligand binding site on the rat adenosine A_{2a} receptor. *Eur. J. Pharmacol.–Mol. Pharm. Sect.* **268**: 95–104.

Jen T, Kaiser C. (1977) Adrenergic agents. 5. Conformational analysis of 1-alkylamino-3-aryloxy-2-propanols by proton magnetic resonance studies. Implications relating to the steric requirements of adrenoceptors. *J. Med. Chem.* **20**: 693–698.

Kobilka B. (1992) Adrenergic receptors as models for G protein-coupled receptors. *Ann. Rev. Neurosci.* **15**: 87–114.

Lewell XQ. (1992) A model of the adrenergic beta-2 receptor and binding sites for agonist and antagonist. *Drug Design Discovery,* **9**: 29–48.

Morris TH, Kaumann AJ. (1984) Different steric characteristics of β_1- and β_2-adrenoceptors. *Naunyn-Schmiedeberg's Arch. Pharmacol.* **327**: 176–179.

Ostrowski J, Kjelsberg MA, Caron MG, Lefkowitz RJ. (1992) Mutagenesis of the β_2-adrenergic receptor: how structure elucidates function. *Ann. Rev. Pharmacol. Toxicol.* **32**: 167–183.

Rekker RF, De Kort HM. (1979) The hydrophobic fragmental constant; an extension to a 1000 data point set. *Eur. J. Med. Chem.–Chim. Ther.* **14**: 478–488.

Strader CD, Sigal IS, Register RB, Candelore MR, Rands E, Dixon RAF. (1987) Identification of residues required for ligand binding to the β-adrenergic receptor. *Proc. Natl Acad. Sci. USA,* **84**: 4384–4388.

Strader CD, Candelore MR, Hill WS, Sigal IS, Dixon RAF. (1989) Identification of two serine residues involved in agonist activation of the β-adrenergic receptor. *J. Biol. Chem.* **264**: 13572–13578.

Strosberg AD, Camoin L, Blin N, Maigret B. (1993) In receptors coupled to GTP-binding proteins, ligand binding and G-protein activation is a multistep dynamic process. *Drug Design Discovery,* **9**: 199–211.

Suryanarayana S, Kobilka BK. (1993) Amino acid substitutions at position 312 in the seventh hydrophobic segment of the β_2-adrenergic receptor modify ligand-binding specificity. *Mol. Pharmacol.* **44**: 111–114.

Chapter 9

Three-dimensional models for β-adrenoceptor–ligand complexes

Terry P. Lybrand

1. Introduction

Integral membrane proteins play a key role in signal transduction across membranes. The G protein-coupled receptor (GPCR) superfamily represents a physiologically and pharmacologically important class of membrane-bound receptor proteins (Savarese and Fraser, 1992; Strader, *et al.*, 1994; Watson and Arkinstall, 1994). At least 1000 proteins in organisms ranging from *Drosophila* to humans have been identified as members of this family. With the exception of the rhodopsins, the adrenergic neurotransmitter receptors are probably the best characterized GPCRs (Kobilka, 1992).

Numerous three-dimensional models have been constructed for adrenergic and other GPCRs (Cronet, *et al.*, 1993; Dahl *et al.*, 1991; Edvardsen *et al.*, 1992; Findlay and Eliopoulos, 1990; IJzerman *et al.*, 1992; Lewell, 1992; Nordvall and Hacksell, 1993; Trumpp-Kallmeyer *et al.*, 1992; Yamamoto *et al.*, 1993), and most have been based on the electron diffraction structure of bacteriorhodopsin (Henderson *et al.*, 1990), although there is no sequence homology observed between bacterio-rhodopsin and the GPCRs. The rationale for use of bacteriorhodopsin as a template structure for model building invokes the general functional homology between bacteriorhodopsin and rhodopsins from higher organisms, the possibility of a shared structural fold, and the fact that the rhodopsins are members of the GPCR family. Two different hypotheses have been proposed to establish homology between bacteriorhodopsin and GPCRs (and thus provide justification for the use of a bacteriorhodopsin template in modelling studies), while explaining the apparent lack of sequence homology. One hypothesis suggests that a gene duplication event for a primordial three- or four-helix bundle protein led to the GPCRs and all *Halobacteria* retinal proteins (Taylor and Agarwal, 1993). A second hypothesis proposes that exon shuffling events led to an effective rearrangement of helices in the bundle, such that there is no observable collinear sequence homology (Pardo *et al.*, 1992). A detailed sequence comparison of all Halobacteria retinal proteins and a number of GPCRs refutes both these hypotheses, however (Soppa, 1994). Soppa concludes that data to support either gene duplication or exon shuffling hypotheses are favourable only when individual proteins (or a small number of proteins) are analysed; when the entire Halobacteria retinal protein family is examined simultaneously with a large number of GPCRs, no data exist to support either hypothesis, or any sequence homology between the two families.

Some modelling studies have raised additional questions regarding the validity of bacteriorhodopsin as a reference structure, on the basis of alternate sequence alignments or modelled ligand–receptor complexes (MaloneyHuss and Lybrand, 1992; Zhang and Weinstein, 1993). In particular, these studies suggest that the helix bundle arrangement may be quite different from that observed in bacteriorhodopsin. A preliminary low resolution electron diffraction structure for bovine rhodopsin (Schertler *et al.*, 1993) suggests that there are some differences in the transmembrane helix bundle packing arrangements.

In the light of these complications and challenges for three-dimensional model construction of GPCRs, we discuss here some modelling strategies and protocols and describe results for an exercise in three-dimensional model construction for β-adrenoceptors.

2. Methodology

Our modelling strategy relies crucially on the availability of solid experimental data for guidance. Experimental data are used to impose constraints (i.e., to define bounds or limits for certain model details). The ideal experimental 'data' is a three-dimensional structure for a highly homologous protein, thus enabling the application of standard homology model building procedures (Blundell, 1991). However, there are no proteins of known structure that are clearly suitable as an homology reference. A true *de novo* prediction of three-dimensional structure for GPCRs is not feasible at present, either. Therefore, some assumptions must be made regarding three-dimensional structural details to limit model choices.

It is preferable to use solid experimental data to define model limitations, but sequence analysis methods may also provide useful clues. For example, multiple sequence alignments and comparisons for many GPCRs, along with hydrophobicity and/or secondary structure analyses, suggest that GPCRs contain seven transmembrane segments that are probably α-helices. These predictions are supported by peptide mapping experiments using anti-peptide antibodies (Wang *et al.*, 1989). Residue substitution patterns observed in multiple sequence alignments also provide strong clues regarding external (i.e. membrane-exposed) versus internal residues. External residues typically exhibit great variability, with the sole restriction being that only hydrophobic residues are tolerated. Internal residues tend to be more highly conserved in subfamilies such as the adrenoceptors, and most polar residues in the transmembrane domain are sequestered in the interior. Residues that are highly conserved in all GPCRs are thought to play a key role as structural elements in these proteins. Residues that are conserved within a receptor subgroup, such as the β-adrenoceptors, but unconserved in other GPCRs, are thought to serve in a specific functional role, such as ligand binding. Alignment of transmembrane sequences, along with peptide antibody mapping experiments, can define the size of interhelical loops. Since some loops in many GPCRs are rather short, they impose topological constraints in the helix packing arrangements (e.g. an extremely short loop connecting two helices forces those two helices to pack next to each other.

These types of constraints permit significant restrictions on three-dimensional model possibilities. Often, there may be only a limited number of seven-helix bundle arrangements that are consistent with these constraints. When the receptor binds small, rigid ligands, further constraints on model options may be possible. Only those helix bundle arrangements that satisfy constraints derived from peptide mapping, disulphide crosslinks, sequence conservation patterns, and also exhibit sensible ligand binding pockets need be considered. In the case of well-characterized GPCRs such as the β-adrenoceptors, there are sufficient reliable constraint data to limit the number of model options to a manageable collection, perhaps only a few dozen or so.

Once model possibilities are limited to a manageable number, additional data must be utilized to select the most viable candidates. A wealth of site-directed mutagenesis data are available for many GPCRs, and this information has been used routinely to select and refine models. Unfortunately, much site-directed mutagenesis data may be of questionable usefulness in the construction and evaluation of GPCR models at this point. Even in cases of proteins with well-known three-dimensional structures, it has often been observed that such data can be confusing, or even misleading. Given the considerable uncertainty regarding nearly all aspects of GPCR three-dimensional structural models, site-directed mutagenesis data must be used quite cautiously. When a particular mutation alters ligand binding affinity by 1000-fold or more, it is reasonable to assume that the residue plays a major (direct) role in ligand interaction. Likewise, a non-conservative mutation that produces no effect on ligand binding is unlikely to be located in the ligand binding pocket. However, attempts to probe ligand binding sites with site-directed mutagenesis yield many cases where ligand binding affinity is modified by factors of tenfold or less. It is quite possible that these subtle ligand binding changes are due to complicated indirect effects, such as minor to modest conformation changes away from the binding pocket. In these cases, the mutagenesis data provide no useful constraints for model construction.

Since single site mutation data may often be of limited use for guiding model construction, other data are necessary. Double mutations that restore activity disrupted by prior single mutations have long been viewed as evidence for probable residue interactions in proteins (e.g., if a single mutation disrupts some activity or property, but a second subsequent mutation restores normal activity, it is generally assumed that the second mutation plays a direct compensatory role to changes introduced by the first mutation). Though it is possible to imagine cases where the effect is an indirect one, it has typically been observed that double site mutant revertants (i.e. double mutations that restore normal biological activity) involve residues that interact directly. This concept of double site revertants may also be extended to ligand–protein interactions in some cases. For a number of GPCRs, it has been shown that site-directed mutagenesis effects may be mimicked by functional modification of ligands (Strader *et al.*, 1989). In some cases, it has also been shown that a disruptive mutation can be compensated by a suitable ligand modification (Strader *et al.*, 1991). Such data argue much more strongly for direct specific interactions between two protein residues or a residue and a ligand than simple mutation data.

Beyond double site revertant data, other types of experimental data may also be extremely useful for three-dimensional model construction efforts. Covalent labelling of receptors with photoaffinity or other covalent labelling ligands can be a powerful tool to characterize ligand binding pockets, provided the affinity labels are not too promiscuous. Selective covalent modification of certain residues such as cysteines with spectroscopically active functional groups can provide detailed structural and dynamical information. For example, modification of cysteines with spin labels or unique fluorophores can enable detailed characterization of inter-residue distances and molecular environments. Systematic introduction of cysteine mutations, coupled with cross-linking and spin-labelling experiments, has been a powerful tool for structural characterization of several membrane proteins (Altenbach *et al.*, 1990; Lee *et al.*, 1994; Steinhoff *et al.*, 1994).

3. Application to adrenoceptor models

We have endeavoured to use reliable experimental data in construction of adrenoceptor models. As the β-adrenoceptors are probably the best characterized of the ligand-activated GPCRs, we have focused primarily on this group of receptors. Multiple sequence alignment of all β-adrenoceptors, along with other assorted GPCRs, reveal most expected patterns. Highly conserved residues, such as prolines in transmembrane (TM) helices 4, 5, 6, and 7, the aspartate-arginine-tyrosine tripeptide motif near the carboxy terminus of TM-3, an aspartic acid (Asp-79 in β_2-receptor) in TM-2, an asparagine (Asn-322 in β_2-receptor) in TM-7, and several other key residues are all present in β-adrenoceptors. Key landmark residues such as these make it rather easy to align the β-receptors with other diverse GPCRs. These alignments also reveal patterns of other highly conserved residues unique to β-receptors and provide important clues regarding orientation of helices with respect to the lipid environment, as discussed above.

There is also a great deal of experimental data on ligand binding for β-adrenoceptors. While much of these data are derived from single mutation experiments, and must be used with caution for the reasons outlined above, other ligand binding data are somewhat more definitive. A highly conserved aspartic acid in TM-3 is proposed to serve as the counter-ion for the protonated amine functional group present in adrenergic ligands (*Figure 1*). Mutation of this aspartate (Asp-113 in β_2-receptor) to asparagine or other neutral residues reduces substantially agonist and antagonist binding (Strader *et al.*, 1988). Further evidence for this interaction is provided by an interesting 'revertant' study. When Asp-113 is mutated to serine, traditional adrenergic ligands do not bind well or at all. However, when an adrenergic ligand is chemically modified such that the amine functional group is replaced by an ester or ketone, the Asp-113-Ser mutant receptor binds this ligand well and is able to transduce a normal agonist signal (Strader *et al.*, 1991). This experiment is roughly analogous to double site revertants discussed above. Similar data are available for a few other key residues in β-adrenoceptors. Multiple sequence alignments reveal two highly conserved serines in TM-5 of all GPCRs that bind catechol ligands. Mutation of either serine to alanine reduces agonist binding affinity; mutation of both serines also yields a receptor incapable of normal signal transduction (Strader *et al.*, 1989). The altered ligand binding properties of these Ser

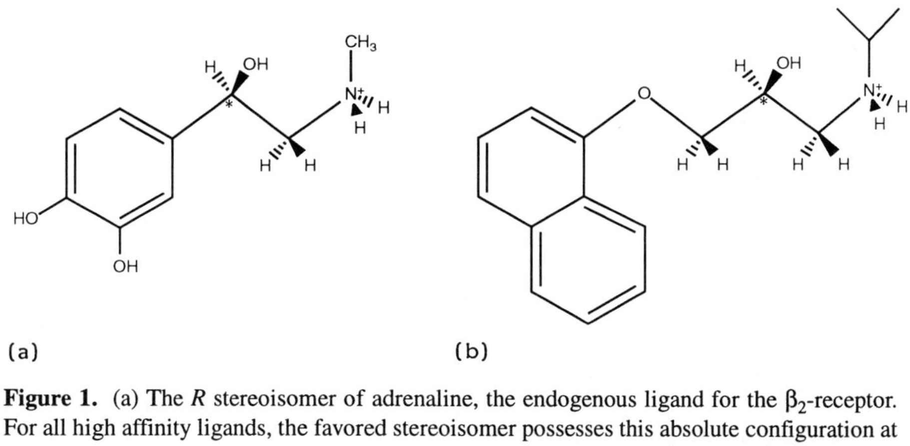

Figure 1. (a) The *R* stereoisomer of adrenaline, the endogenous ligand for the β_2-receptor. For all high affinity ligands, the favored stereoisomer possesses this absolute configuration at the chiral side chain carbon. (b) The *S* stereoisomer of propranolol, a β-adrenergic antagonist. Note that the *S* stereoisomer of propranolol, the preferred stereoisomer, possesses the same absolute configuration at the side chain chiral carbon as the *R* isomer of adrenaline.

to Ala mutants are mimicked rather well in wild-type receptors by removal of the corresponding hydroxyl groups from the catechol ring of β-agonists, suggesting strongly that Ser-204 interacts with the *m*-hydroxyl group and Ser-207 interacts with the *p*-hydroxyl group (Strader *et al.*, 1989). While there are many other site-directed mutagenesis studies that implicate certain receptor residues in ligand interactions, the results are much less definitive. Of particular frustration has been the inability thus far to identify the receptor residue(s) responsible for interaction with the hydroxyl group attached to the chiral carbon of β-adrenergic ligands. It seems reasonable to assume that this interaction dictates stereoselectivity in β-receptors, and conclusive experimental data to define these receptor residues would provide important additional constraints for model building exercises.

With these specific experimental data as initial constraints, and clues from multiple sequence alignments, we have constructed a number of detailed three-dimensional models for the β_2-adrenoceptor. The only assumption made in all our modelling exercises is that the β-receptors (and other GPCRs) contain a seven-helix bundle that spans the lipid bilayer. Given the results of sequence analyses and multiple sequence alignments, together with peptide antibody mapping experiments and other biophysical probes, this assumption seems well justified. In *de novo* model construction, it is then necessary to decide how the seven helices pack together to form a bundle. In the absence of any topological constraints, there are in principle $7! = 5040$ different packing arrangements for seven helices in a closed ring-like pattern.

Clues regarding the orientation of each helix *vis à vis* the lipid bilayer can be deduced from the sequence alignment results, as described previously. In the β-adrenoceptors, as in many other GPCRs, some of the interconnecting loops between helical regions are rather short (as short as 8–9 residues in some cases). These short loops act as tethers that force certain helices to pack next to each other, provided loop backbone torsion angles are required to adopt energetically reasonable conformations. Additionally, there are two cysteines highly conserved in most GPCRs (one in the extracellular loop between TM-2 and TM-3 and the other in the extracellular loop between TM-4 and TM-5). These cysteines are believed to form a disulphide bridge in most GPCRs, as mutation of either generally yields receptors that have diminished ligand binding and/or signal transduction capabilities. The combination of some short interconnecting loops and other topological constraints like disulphide bonds between loops, restrict the number of possible arrangements for a seven-helix bundle. For example, the connecting loop between TM-6 and TM-7 in the β_2-receptor is sufficiently short to require that these two helices pack next to each other. The extracellular loop between TM-2 and TM-3 is also relatively short, and the positions of the two conserved cysteines in extracellular loops suggest that TM-3 is probably sandwiched between TM-2 and TM-4 (provided backbone torsion angles in these loop regions are restricted to energetically reasonable values that also permit formation of the proposed disulphide bridge). With these restrictions on helix placement, positions for TM-1 and TM-5 relative to TM-2,3,4 and TM-6,7 clusters can be explored exhaustively, as there are now only 4! = 24 arrangements possible (again, assuming some type of closed ring pattern for the seven-helix bundle and reasonable backbone geometries for connecting loop regions).

Further clues for helical bundle arrangement may be drawn from the ligand binding site characterisation studies discussed above. Viable three-dimensional receptor models must congregate residues involved in direct ligand interactions in a localised region, thus creating a ligand binding pocket. In addition to these experimental and topological constraints, any feasible three-dimensional receptor model must exhibit reasonable physical property characteristics; backbone and side chain torsion angles must adopt energetically reasonable values, side chain packing densities should resemble that of well-characterised proteins, and exposed surface properties should be consistent with the environment (e.g. the exterior of the transmembrane seven-helix bundle should be hydrophobic for compatibility with the lipid bilayer).

Our three-dimensional model generation procedures have been described previously in detail (MaloneyHuss and Lybrand, 1992), so only a short synopsis will be given here. Construction of possible seven-helix bundles is accomplished using interactive molecular graphics modelling programs, although computer-automated search procedures are certainly possible. For each seven-helix bundle that satisfies the experimental constraints outlined above, interconnecting loops are then attached to ensure that the bundle arrangement is compatible with conformationally reasonable loop geometries. The loop geometry restraints arise not only from allowable backbone and side chain torsion angle values, but also from disulphide crosslinks and any other topological constraints that might be present. While numerous techniques are available to generate loop structures for three-dimensional protein models (e.g. search of libraries composed of loop structures from protein crystal structures, residue by residue build-up based on Monte Carlo or systematic conformational search methods), some of the longer loops and loops with disulphide crosslinks cannot be constructed easily with these methods. Instead, we have used short, low temperature molecular dynamics simulations to generate loops. Each loop peptide is first generated in an extended conformation. One end of the peptide is then attached to the end of a transmembrane helix with a *trans* peptide bond using interactive graphics. Weak harmonic constraints are applied to pull the free end of the peptide into position to form a *trans* peptide bond with the second target helix. Generally, only 1–2 picoseconds of constrained molecular dynamics at 20–50 K is necessary to ligate the loop segments on to the helix bundle. These low temperature molecular dynamics simulations efficiently close loop structures without introduction of strain in the peptide backbone. Loops with crosslinks are generated simultaneously; one end of each peptide is attached to an appropriate helix, and sufficient constraints are applied to close both loop structures and form the disulphide crosslink in one process. The loop conformations generated with this procedure are not unique, or even the lowest energy conformations possible, but they are energetically reasonable loop structures and can thus be helpful in assessing helix bundle models. As discussed above, some helix bundle arrangements require loop conformations that are badly strained, suggesting that the helix bundle model itself is improbable.

After helix bundle models are created that satisfy experimental and topological constraints, each is evaluated for reasonable physical properties. Standard Ramachandran plots are used to assess backbone geometry. Side chain torsion angles are compared with rotamer libraries constructed by analysis of high resolution protein crystal structures (Dunbrack and Karplus, 1993; Ponder and Richards, 1987). Side chain packing densities are computed using the QPACK program (Gregoret and Cohen, 1990). Simple geometric tests like these have proven quite powerful in evaluation of both crystal and model protein structures, and permit rapid elimination of physically unreasonable models. Once three-dimensional models are identified that satisfy all input constraints and are also physically reasonable, they are then challenged with additional experimental data not included in the initial modelling constraints to further assess quality. Since numerous high affinity ligands are known for β-adrenoceptors, we have docked ~30 agonists and antagonists of varying sizes and structural classes in our receptor models (Kontoyianni and Lybrand, submitted). A good three-dimensional β-receptor model should accommodate low energy conformations for a wide range of ligands with at most only minor structural modifications. All these docked ligand complexes must also satisfy constraints imposed in the original model construction and possess sensible receptor interactions for all ligand functional groups (e.g. hydrogen bonds, aromatic and aliphatic interactions, etc.).

Two general model structures have emerged from our β-adrenoceptor construction exercises. One model exhibits a sequential, counterclockwise arrangement of the seven-helix bundle in a pattern vaguely similar to bacteriorhodopsin (*Figure 2a*). This model shares many features reported for other β-receptor models. The other model possesses a clockwise helix bundle arrangement (*Figure 2b*). Both models satisfy the constraint data and physical property assessment tests described above equally well. The most noticeable difference in the two models concerns implications for stereoselective ligand binding. When ligands are docked to form the key side chain interactions described above, the hydroxyl group attached to the ligand chiral carbon (i.e. the stereoselective attachment site for adrenergic ligands) is oriented toward TM-4 in the counterclockwise model, and TM-7 in the clockwise model (*Figure 2*). There are three residues in TM-4 that are good candidates to form hydrogen bond interactions with this ligand hydroxyl group, Ser-161, Thr-164, and Ser-165. Mutation of Ser-161 to alanine has no effect on ligand binding

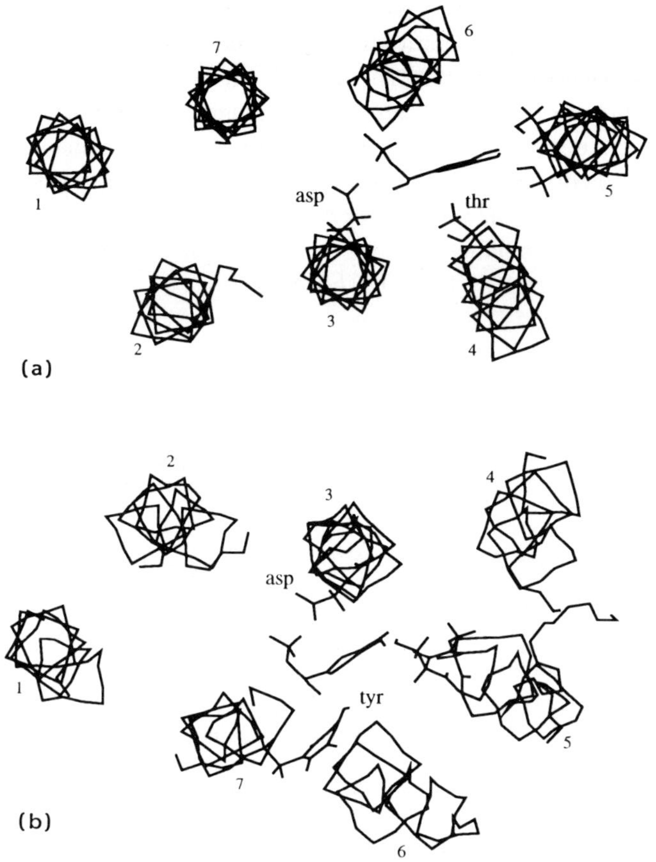

Figure 2. (a) A view of the counterclockwise β_2 adrenoceptor model from the extracellular side. Loops have been clipped away to thin slice view of the ligand binding pocket. Adrenaline is pictured in the interior of the helix bundle. Aspartate 113 from transmembrane helix 3 and Thr-164 (a possible stereoselective hydrogen bond partner) are labelled. Each helix is numbered. (b) A view of the clockwise β_2 receptor model. Labelling is as specified for panel A, except that the putative hydrogen bonding residue in this model is Tyr-316 in transmembrane helix 7.

(Strader *et al.*, 1989), so this residue is probably not involved in direct ligand interactions. Recent studies have reported that mutation of Thr-164 slightly reduces binding affinity for some ligands, but has no effect on other, structurally similar ligands (Green *et al.*, 1993). Furthermore, multiple sequence alignments show that this threonine is conserved only in β_2-receptors, but not β_1- or β_3-receptors. It is difficult to interpret these complex, and somewhat contradictory, observations, but the current data do not seem to provide strong evidence for Thr-164 as the key stereoselective binding site. Mutation of Ser-165 to alanine yields a receptor that appears not to fold properly (Strader, *et al.* 1989). These data would seem to imply that Ser-165 is important for proper protein folding. Important structural residues are generally not observed to participate in key, specific ligand interactions in proteins of known structure, so the role of Ser-165 as the major stereoselective binding site is also suspect. Attempts to identify other residues in TM-4 that may be involved in specific ligand binding interactions have not been successful (Dixon *et al.*, 1989). There are also several interesting residues in TM-7 that are good candidates for the stereoselective binding site in the clockwise model. Ser-319 is well positioned to form a stereoselective hydrogen bond with ligands. Mutation of Ser-319 to alanine produces a small (10-fold) reduction in agonist binding affinity, but does not seem to affect antagonist binding (Strader *et al.*, 1989). It also appears that the Ser-319 to Ala mutation lowers binding affinity equally for both *R* and *S* stereoisomers of agonists (CD Strader, personal communication). Hence, this residue is unlikely to function as the principal stereoselective binding site. Tyr-316 is also an interesting possibility. After limited molecular dynamics refinement of ligand complexes in the clockwise model (Kontoyianni and Lybrand, 1993), most ligands form a stereoselective hydrogen bond with Tyr-316, at the expense of any other hydrogen bonding interaction that may have been imposed in ligand docking exercises. A Tyr-316 to Ala mutant has been constructed, although no ligand binding measurements were reported for this mutant receptor (Strader *et al.*, 1989). However, this is not a particularly conservative mutation, and it would be quite interesting to examine the effect of a tyrosine to phenylalanine mutation at this position. A few other residues are less likely candidates for the stereoselective binding site. Asn-312 is in the binding pocket region, but is less well positioned to form a good hydrogen bond with the side chain hydroxyl group. We do observe that this asparagine forms hydrogen bonds with the ether oxygen of propanolamine antagonists, such as propranolol, after molecular dynamics structural refinement, and there is some experimental data to suggest this interaction may occur (Suryanarayana and Kobilka, 1993).

In addition to mutagenesis and ligand binding data, biophysical studies such as fluorescence spectroscopy and photoaffinity labelling, provide useful data for evaluation of receptor models. A series of fluorescence quenching studies suggest that carazolol (*Figure 3*), a high affinity β_2 antagonist, binds with its chromophore in a hydrophobic pocket that is at least 11–12 Å below the extracellular membrane surface (Tota and Strader, 1990), in good agreement with our placement of the chromophore binding pocket in the model structures. A carazolol derivative modified with a *p*-bromoacetamidobenzyl substituent on the amine group functions as an efficient photoaffinity label and covalently binds the β_2 receptor at a localized peptide fragment in TM-2. In our models, the photoaffinity label is positioned near Ser-92 and His-93 in TM-2, when the ligand docking constraints discussed above are imposed. Characterization of the covalent adducts formed with the carazolol

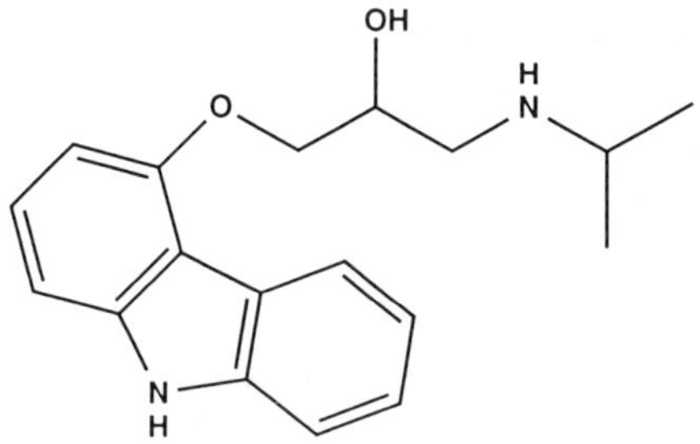

Figure 3. Carazolol, a high affinity β_2 receptor antagonist used often as a fluorescent probe in biophysical studies.

derivative indicates that indeed Ser-92 and/or His-93 are the targets for the crosslinking reaction (Dohlman *et al.*, 1988). Again, the clockwise model seems to fit better with this photoaffinity labelling data, but plausible crosslinked structures are also conceivable in the counterclockwise model. Clearly, at present there is no experimental data from ligand binding or crosslinking studies that can identify cleanly and definitively the stereoselective binding residue(s), and thus enable us to choose the clockwise or counterclockwise models. The numerous ligands we have docked in the model receptors all seem to fit much better in the clockwise model, but these are qualitative evaluations that may be sensitive to modest adjustments in the receptor structure.

Recent mutagenesis studies have provided new data that may allow better model discrimination. It has been observed for some time in multiple sequence alignments that nearly all GPCRs have a conserved aspartic acid in TM-2 and an asparagine in TM-7. Most GPCRs that do not conserve these two positions have them interchanged. The gonadotropin releasing hormone receptor is one example of a GPCR with these two positions reversed from the normal profile. Zhou *et al.* (1994) have performed a number of mutations to explore the effect of these residues. When either position is mutated to the companion residue (Asn to Asp in TM-2 or Asp to Asn in TM-7), the receptor loses ability to bind ligands and transduce signals. When a double mutation is performed (generating the normal GPCR profile with Asp in TM-2 and Asn in TM-7), ligand binding affinity and signal transduction capability are recovered to essentially wild type levels (Zhou *et al.*, 1994). All these data strongly imply a direct interaction between the two residues. Interestingly, this Asp/Asn interaction pair is observed in our clockwise model after molecular dynamics refinement of the structure. This interaction was not imposed during manual model construction, and no constraints were applied during molecular dynamics refinement to generate this interaction.

In contrast, our counterclockwise model does not exhibit this Asp/Asn interaction; the helix orientations dictated by ligand binding constraints and conserved versus hydrophobic faces of helices yield a model that places Asp-79 from TM-2 and Asn-322 from TM-7 in nearly opposite orientations to each other. In our counterclockwise model, there is no obvious way to reorient these two residues toward each other that does not expose numerous conserved and polar residues to lipid bilayer or distort the helices dramatically. These are the first experimental data not directly related to ligand binding studies that seem to fit the clockwise model much better than the counterclockwise model

4. Discussion

This brief summary of our β-adrenoceptor construction illustrates some of the capabilities, and limitations, of current modelling techniques and strategies. It is clear from these modelling exercises that experimental data to be used for constraints must be chosen carefully. Tenuous or ambiguous constraint data will bias three-dimensional models in unreasonable ways, or more likely, present internal inconsistencies that cannot be resolved. Data from point mutation studies are often of questionable usefulness in model construction, although these data may be helpful in subsequent evaluation of three-dimensional model structures. Fortunately, it is relatively easy to define 'good' model constraint data. Results from direct biophysical probes of receptor structure (e.g. fluorescence spectroscopic or spin labelling studies), double site revertant mutations, or single mutations that produce dramatic effects on ligand binding without obvious impact on general receptor structure, may often be used to good advantage in receptor model building.

Our use of a *de novo* model construction protocol has led us to explore three-dimensional models that probably would not be considered under other circumstances, such as clockwise seven-helix bundle models. The results of these exercises show clearly that several physically sensible three-dimensional models are possible for the β-adrenoceptors, each of which is consistent with basic biochemical and structural data. There is not yet sufficient experimental data to define a single, best structural model. In other words, model construction using experimental data as constraints is still a severely underdetermined problem for GPCRs. Nonetheless, it is encouraging that many possible three-dimensional models can be eliminated based on inconsistency with experimental data or topological constraints.

Inability to define a single, optimal model structure does not diminish the usefulness of the modelling studies, however. We use these three-dimensional models in much the same manner as many other researchers; namely, as a tool to stimulate hypotheses for receptor structure/function properties and to design new experiments. Our model building exercises have helped us understand possible molecular details and implications of numerous experimental studies. The three-dimensional models also reveal details of tentative receptor structure and receptor–ligand interactions (e.g. Tyr-316 direct interactions with ligand side chain hydroxyl group) that have not been considered previously. Thus, the models can help suggest specific new experimental studies that will yield additional information about these receptors. In turn, the models can be challenged with data generated from the new experiments, and then modified or discarded, as appropriate. An iterative process of model building and experimental testing should make it possible eventually to construct quite realistic three-dimensional models.

There is also interest in the use of these three-dimensional receptor models as tools in rational drug design. As mentioned previously, many GPCRs are therapeutic targets of great potential significance. If three-dimensional models for GPCRs attain sufficient accuracy, it may be possible to use standard ligand docking algorithms or other modelling tools to help design new drugs. There are some documented cases where computer-based ligand docking searches, conducted using model receptor sites, have yielded useful lead compounds for therapeutic development (Ring *et al.*, 1993). It is not clear that current models for GPCRs are adequate for this purpose at present, but it is certainly a reasonable goal. Use of modelled receptor–ligand complexes to help optimize lead compounds is another possible application. However, lead compound optimization will probably require greater structural detail accuracy than current *de novo* modelling protocols can achieve in the near future. Even when high resolution crystal structures are available for receptor site targets, lead compound optimization with computational methods has proven difficult.

5. Conclusions

Our results show that a *de novo* model building protocol can be used to generate physically plausible three-dimensional structures for GPCRs. These three-dimensional models can aid in analysis, interpretation, and even design, of experimental studies. When coupled with appropriate experimental techniques, model building exercises can provide additional insights for GPCR structural and functional properties. There is also a reasonable possibility that receptor models may soon be useful in some rational drug design applications. Given the enormous technical challenges that complicate direct structural characterization of these important integral membrane proteins, model building exercises are likely to play a meaningful role in the study of GPCRs in the future.

References

Altenbach C, Marti T, Khorana HG, Hubbell WL. (1990) Transmembrane protein structure: spin labeling of bacteriorhodopsin mutants. *Science,* **248**: 1088–1092.

Blundell, TL. (1991) Comparative analysis of protein three-dimensional structures and an approach to the inverse folding problem. *Ciba Found. Symp.* **161**: 28–36.

Cronet P, Sander C, Vriend G. (1993) Modeling of transmembrane seven-helix bundles. *Protein Eng.* **6**: 59–64.

Dahl SG, Edvardsen Ø, Sylte I. (1991) Molecular dynamics of dopamine at the D_2 receptor. *Proc. Natl Acad. Sci. USA*, **88**: 8111–8115.

Dixon RAF, Hill WS, Candelore MR, Rands E, Diehl RE, Marshall MS, Sigal IS, Strader CD. (1989) Genetic analysis of the molecular basis for β-adrenoceptor subtype specificity. *Proteins: Struct. Funct. Genet.* **6**: 267–274.

Dohlman HG, Caron MG, Strader CD, Amlaiky N, Lefkowitz RJ. (1988) Identification and sequence of a binding site peptide of the β_2-adrenergic receptor. *Biochemistry*, **27**: 1813–1817.

Dunbrack RL, Karplus M. (1993) Backbone-dependent rotamer library for proteins. Application to side chain prediction. *J. Mol. Biol.* **230**: 543–574.

Edvardsen Ø, Sylte I, Dahl SG. (1992) Molecular dynamics of serotonin and ritanserin interacting with the 5-HT2 receptor. *Mol. Brain. Res.* **14**: 166–178.

Findlay J, Eliopoulos E. (1990) Three-dimensional modeling of G protein-linked receptors. *Trans Pharm. Sci.* **11**: 492–499.

Green SA, Cole G, Jacinto M, Innis M, Liggett SB. (1993) A polymorphism of the human β_2-adrenergic receptor within the fourth transmembrane domain alters ligand binding and functional properties of the receptor. *J. Biol. Chem.* **268**: 23116–23121.

Gregoret LM, Cohen FE. (1990) Novel method for the rapid evaluation of packing in protein structures. *J. Mol. Biol.* **211**: 959-974.

Henderson R, Baldwin JM, Ceska TA, Zemlin F, Beckmann E, Downing KH. (1990) Model for the structure of bacteriorhodposin based on high-resolution electron cryo-microscopy. *J. Mol. Biol.* **213**: 899–929.

IJzerman AP, van Galen PJM, Jacobson KA. (1992) Molecular modeling of adenosine receptors. I. The ligand binding site on the A_1 receptor. *Drug Design Discovery,* **9**: 49-67.

Kobilka B. (1992) Adrenergic receptors as models for G protein-coupled receptors. *Ann. Rev. Neurosci.* **15**: 87–114.

Kontoyianni M, Lybrand TP. (1993) Computer modeling studies of G protein-coupled receptors. *Med. Chem. Res.* **3**: 407–418.

Lee GF, Burrows GG, Lebert MR, Dutton DP, Hazelbauer GL. (1994) Deducing the organization of a transmembrane domain by disulfide cross-linking. The bacterial chemoreceptor Trg. *J. Biol. Chem.* **269**: 29920–29927.

Lewell XQA. (1992) A model of the adrenergic beta-2 receptor and binding sites for agonist and antagonist. *Drug Design Discovery,* **9**: 29–48.

MaloneyHuss K, Lybrand TP. (1992) Three-dimensional structure for the b_2 adrenergic receptor protein based on computer modeling studies. *J. Mol. Biol.* **225**: 859–871.

Nordvall G, Hacksell U. (1993) Binding-site modeling of the muscarinic m1 receptor: A combination of homology-based and indirect approaches. *J. Med. Chem.* **36**: 967–976.

Pardo L, Ballesteros JA, Osman R, Weinstein H. (1992) On the use of the transmembrane domain of bacteriorhodopsin as a template for modeling the three-dimensional structure of guanine nucleotide-binding regulatory protein-coupled receptors. *Proc. Natl Acad. Sci. USA*, **89**: 4009–4012.

Ponder JW, Richards FM. (1987) Tertiary templates for proteins; use of packing criteria in the enumeration of allowed sequences for different structural classes. *J. Mol. Biol.* **193**: 775-791.

Ring CS, Sun E, McKerrow JH, Lee GK, Rosenthal PJ, Kuntz ID, Cohen FE. (1993) Structure-based inhibitor design by using protein models for the development of antiparasitic agents. *Proc. Natl. Acad. Sci. USA*, **90**: 3583–3587.

Savarese TM, Fraser CM. (1992) In vitro mutagenesis and the search for structure–function relationships among G protein-coupled receptors. *Biochem. J.* **283**: 1–19.

Schertler GFX, Villa C, Henderson R. (1993) Projection structure of rhodopsin. *Nature*, **362**: 770-772

Soppa J. (1994) Two hypotheses — one answer: Sequence comparison does not support an evolutionary link between halobacterial retinal proteins including bacteriorhodopsin and eukaryotic G-protein-coupled receptors. *FEBS Lett.* **342**: 7–11.

Steinhoff HJ, Mollaaghabata R, Altenbach C, Hideg K, Krebs M, Khorana HG, Hubbell WL. (1994) Time-resolved detection of structural changes during the photocycle of spin-labeled bacteriorhodopsin. *Science*, **266**: 105–107.

Strader CD, Sigal IS, Candelore MR, Rands E, Hill WS, Dixon RAF. (1988) Conserved aspartic acid residues 79 and 113 of the β-adrenergic have different roles in receptor function. *J. Biol. Chem.* **263**: 10267–10271.

Strader CD, Candelore MR, Hill WS, Sigal IS, Dixon RAF. (1989) Identification of two serine residues involved in agonist activation and the β-adrenergic receptor. *J. Biol. Chem.* **264**: 13572–13578.

Strader CD, Gaffney T, Sugg EE, Candelore MR, Keys R, Patchett AA, Dixon RAF. (1991) Allele-specific activation of genetically engineered receptors. *J. Biol. Chem.* **266**: 5–8.

Strader CD, Fong TM, Tota MR, Underwood D, Dixon RAF. (1994) Structure and function of G protein-coupled receptors. *Annu. Rev. Biochem.* **60**: 101-132.

Suryanarayana S, Kobilka BK. (1993) Amino acid substitutions at position 312 in the seventh hydrophobic segment of the β_2-adrenergic receptor modify ligand-binding specificity. *Mol. Pharmacol.* **44:** 111–114.

Taylor EW, Agarwal A. (1993) Sequence homology between bacteriorhodopsin and G-protein coupled receptors: exon shuffling or evolution by duplication? *FEBS Lett.* **325**: 161–166.

Tota MR, Strader CD. (1990) Characterization of the binding domain of the b-adrenoceptor with the fluorescent antagonist carazalol. *J. Biol. Chem.* **265**: 16891–16897.

Trump-Kallmeyer S, Hoklack J, Bruinvels A, Hibert M. (1992) Modeling of G-protein-coupled receptors: application to dopamine, adrenaline, serotonin, acetylcholine, and mammalian opsin receptors. *J. Med. Chem.* **35**: 3448–3462.

Wang H, Lipfert L, Malbon CC, Bahouth S. (1989) Site-directed antibodies define the topography of the β-adrenergic receptor. *J. Biol. Chem.* **264**: 14424–14431.

Watson S, Arkinstall S. (1994) *The G-Protein Linked Receptor Factsbook.* New York: Academic Press.

Yamamoto Y, Kamiya K, Terao S. (1993) Modeling of human thromboxane A_2 receptor and analysis of the ligand-receptor interaction. *J. Med. Chem.* **36**: 820–825.

Zhang D, Weinstein H. (1993) Signal transduction by a 5-HT_2 receptor: A mechanistic hypothesis from molecular dynamics simulations of the three-dimensional model of the receptor complexes to ligands. *J. Med. Chem.* **36**: 934–938.

Zhou W, Flanagan C, Ballesteros JA, Konvicka K, Davidson JS, Weinstein H, Millar RP, Sealfon SC. (1994) A reciprocal mutation supports helix 2 and helix 7 proximity in the gonadotropin-releasing hormone receptor. *Mol. Pharmacol.* **45:** 165–170.

Chapter 10

Computational tools and results in the construction of G protein-coupled receptor models

Frank E. Blaney and Mike Tennant

1. Introduction

The malfunction of G protein-coupled receptor (GPCR) systems has been linked to a wide number of disease states; hence a better understanding of the structure of the GPCRs is of immense importance to the pharmaceutical industry. For this reason most major companies have invested considerable effort in the construction of three-dimensional models of GPCRs with the hope of using these to gain insight into their binding modes and to use them in the design of novel, potentially therapeutic ligands. Although no structural data from X-ray diffraction or nuclear magnetic resonance spectroscopy is available for the GPCRs, a large amount of information has become available on the sequences of these receptors (Probst *et al.*, 1992). Hydropathicity analysis of these sequences, using methods such as that of Kyte and Dolittle (1982) or Engelman *et al.* (1986), has shown that they most likely consist of seven transmembrane α-helices and it has therefore been suggested that they are structurally homologous to the membrane bound protein, bacteriorhodopsin (bR). Although the sequence homology of the GPCRs with bacteriorhodopsin is very low (approximately 16%), the latter is known to have a similar hydropathicity profile and therefore most of the modelling work reported to date has been based on this structural motif. The publication by Schertler *et al.* (1993) of the first electron diffraction projection map of a GPCR, bovine rhodopsin, has given further credance to this work.

One advantage which the industry has in this modelling work is the availability of ligand binding data on thousands of compounds. In the development of three-dimensional models of GPCRs therefore, an explanation of the quantitative structure–activity relationship (QSAR) for a series of diverse ligands with a given receptor, should be a good validation of a three-dimensional model of this receptor. In our early work in this field, models based loosely on the bacteriorhodopsin packing, template, were manually generated using the widely accepted CHARMm program (Brookes *et al.*, 1983). This required the construction of individual CHARMm scripts for each receptor which often took the better part of a day to do. It soon became clear that to generate the best possible models to explain these QSARs, a method was needed which would (a) allow for the rapid construction of a variety of alternative models based on different helix packings, solvation states, etc. and (b) automate in an appropriate manner, the positioning and evaluation of ligands in the receptors in a variety of energetically favourable docked conformations. It was for these reasons that a program which we have named GPCR_Builder was developed. The importance of such a program to us has increased enormously as a result of the impact of molecular biology which has led to a rapid increase in the number of novel sequences. This chapter describes work which has gone into the development of the first part of the program.

2. Construction of three-dimensional GPCR models

The perceived 'homology' with bacteriorhodopsin has already been noted above and with this fact in mind, three approaches have been used in the modelling of GPCRs. The first is based on the assumption that, as homology with bacteriorhodopsin is so low, it is better to build the proteins *de novo*. This approach has been used by Evardsen *et al.* (1992) for the generation of models of dopamine D_2 and 5-HT_2 receptors. They have assumed a circular antiparallel packing of the helices, which has subsequently been subjected to molecular dynamics in the presence of bound ligands. A more detailed model of the β_2 receptor was generated by MaloneyHuss and Lybrand (1992). This model included all the extracellular and cytoplasmic loops and the N- and C-termini. Rather than minimization techniques, they used rotamer libraries to define the side chain dihedrals. A rather different approach has been used by Zhang and Weinstein (1993) again for the 5-HT_2 receptor. Instead of using the standard consecutive arrangement of the helices, they have based their structure on a previous model of activation of this receptor, together with the results of chemical and chimeric experiments and have come up with a rather different packing arrangement of the seven helices. This model was used to generate a plausible hypothesis on the activation of the receptor by serotonin.

The other main approach to building GPCR models is to base the structure on that of bacteriorhodopsin. Despite the low homology this is not unreasonable, because throughout the field of protein structure, remarkable similarity has been observed in structural motifs between proteins with no primary sequence homology. Some groups, notably those of Trumpp-Kallmeyer *et al.* (1992), Livingstone *et al.* (1992) and Osman (private communication), have maintained that the helices should be built individually before final alignment with the bacteriorhodopsin bundle. The logic behind this is that this method will allow the proline kinks to develop properly in the five helices which contain them. Trump-Kallmeyer *et al.* (1991; 1992) used minimization methods on the helices after they had been constructed with ideal alpha-helical geometry. Osman, on the other hand, used molecular dynamics on the individual helices before final assembly. Lewell (1992) has used the alternative method of mutating the residues of bacteriorhodopsin to the 'corresponding' ones in the β_2-adrenoceptor to build her model of this GPCR. This model has been used to explain much of the known mutation and experimental binding data.

In our early work in this field, we have also used the latter approach in the construction of various neurotransmitter and peptide GPCR models. To get around the problem of non-correspondence of the prolines with bacteriorhodopsin, we set large phi–psi dihedral constraints to restrict the helix conformations during their initial minimization. These were gradually decreased during the course of the minimization and checks were maintained on the phi-psi plot to ensure that the structure did not break up. Several of these models, notably of dopamine D_2 serotonin (5-HT_{2a}) and endothelin were used as the initial templates for the GPCR_Builder program.

3. Methods

In all our work in the field of GPCR modelling, molecular mechanics and dynamics calculations of the protein structures and of the bound ligands were performed using the CHARMm program (Brookes *et al.*, 1983). Geometries for the latter were generated by optimization either with the COSMIC molecular mechanics force field (Vinter *et al.*, 1987) or, where parameters were unavailable, with the semi-empirical molecular orbital program VAMP (Clark *et al.*, 1991), using the AM1 Hamiltonian. Natural orbital charges were used for the docking calculations and these were generated from single point Hartree-Fock calculations on the optimized geometries using the Spartan suite of programs (Wavefunction Inc., Irvine, CA, USA) with a 3-21G* basis set. The results for the CHARMm calculations were visualized using the program, QUANTA (Molecular Simulations Inc., Burlington, MA, USA). All calculations were performed on a Silicon Graphics 380 Powerserver computer and visualized on a Silicon Graphics 4D-310 VGX workstation. The GPCR_Builder program was developed on the latter work-station and was written in Fortran 77 (or C for the text display routines) using the **gl** graphics programming interface.

Origin of the GPCR_Builder Program

The first version of this program was developed for the purpose of rapid construction of CHARMm scripts using template files. Each template consisted of two parts; (a) an alignment data (.adf) file and (b) an associated three-dimensional backbone file for the helix bundle. During the set-up procedure, a 'question and answer' session was carried on via the keyboard which finally resulted in the generation of a CHARMm script file and an associated starting coordinate file. These could then be submitted to an appropriate computer server on the network. As already noted, the templates were chosen from our early models of GPCRs. To make the program more generally useful, this early version was modified primarily to make use of pull-down menus, librarian and dialogue boxes and pseudo-dials. While the use of these templates led to the generation of many useful GPCR models, the general problems associated with the generation of membrane-bound protein models prompted us to develop additional modules which were designed to assist in the construction of alternative novel templates. These are described in greater detail later.

Generation of a three-dimensional GPCR model using the program

The first stage in the construction of a three-dimensional model is of course to read in the sequence. Files are read in as one or (less commonly) three letter codes. A hydropathicity profile calculation is then performed using the methods of Kyte and Dolittle (1982) or Engelman *et al.* (1986) (any other profile can readily be included). The approximate regions containing the seven TM helices are selected either automatically or by using the mouse to pick boxes around the helical regions (*Figure 1*).

The automatic procedure is dependent on the program's ability to find seven lengths of sequence with a continuous calculated hydrophobicity greater than some user-defined threshold. In cases where the interhelical loops are short (which is often the case for example, between TM regions 6 and 7) this procedure can fall down. It is generally preferable therefore to use the manual option. It must be remembered that the program is not attempting to define the actual TM regions at this point. Rather

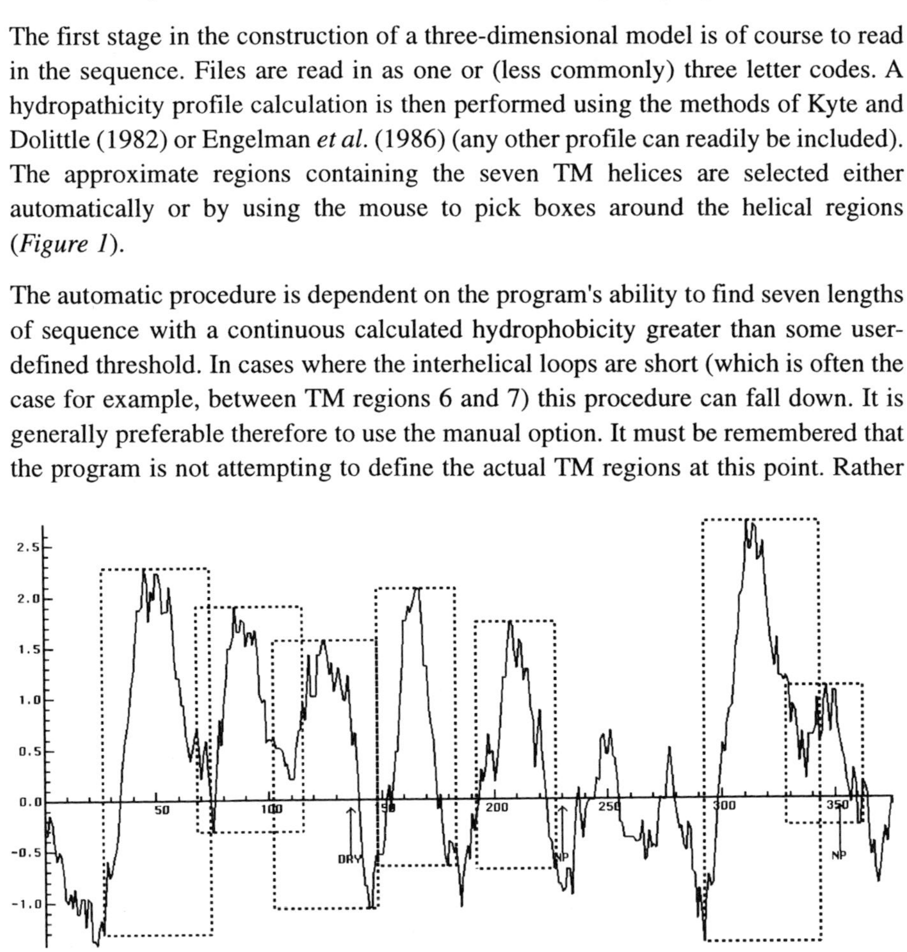

Figure 1. A Kyte–Dolittle plot of a 5-HT_{1d} GPCR sequence with the approximate seven TM regions blocked off. Three DRY and NP positions are indicated for ease of blocking.

it is picking seven regions, each of which must contain the appropriate helix of interest. The actual TM regions used in the model are obtained by alignment with the seven helical sequences in the alignment data file (.adf file) using the 'rule' defined in the file. A typical .adf file is shown in *Figure 2.*

Alignment models currently available include the early models of neurotransmitter receptors such as serotonin and dopamine based on bacteriorhodopsin, Weinstein's 5-HT_{2a} model, some peptide structures again based on bacteriorhodopsin and some

This template is based on the model of the 5-HT_{2c} receptor which was built as a construction of the whole 7 helical bundle aligned to BR and then total minimization. The sequence alignment used was that of Hibert in the Mol. Pharmacol. paper[a]. These lines are simple title lines and end with a blank line.

25 25 22 21 22 25 This line contains the number of residues in each of the template's 7 helices.
WPALSIVVIIIMTIGGNILVIMAVS
FLMSLAIADMLVGLLVMPLSLLAIL
CPVWISLDVLFSTASIMHLCAI
KIAIVWAISIGVSVPIPVIGL These 7 lines are the actual residues in each helix of the template.
LIGSFVAFFIPLTIMVITYFL
VLGIVFFVFLIMWCPFFITNIL
KLMEKLLNVFVWIGYVCSGINPLVY
DATA_DIREC/CATIONIC_GENERAL_BACK_5HT2.CRD This is the 3D template file
Helix1 N-17 V-20
Helix2 S(N)-4 L-5 A-6 D-9 P-18
Helix3 D-25 R-26 Y-27 C-1 W(Y)(F)-4
D-8 T-13 S-15 I-16 L-19
Helix4 W)V)-6 S(A)-13 P(L)(M)(V)-16
Helix4 W(V)-6 S(A)-13 P(L)(M)(V)-16
Helix5 F-8 P-11 Y-19
Helix6 F-9 W-13 P-15 F-16 F-17
Helix7 N-21 P-22 W-12 G-14 Y-25

These lines contain the rules used for alignment. For example Helix 4 normally has a tryptophan (W) at position 6 although this is sometimes a valine (the V in parenthesis) Helix 5 has a phe at position 8, a Pro at 11 and a Tyr at 19 etc. These rules are used to align the corresponding chosen residues in the new sequence.

Figure 2. A typical .adf file. [a]Trumpp-Kallmeyer *et al.* (1991).

newly constructed *de novo* models based on a combination of the recently published electron density map of rhodopsin and some of the analysis modules discussed below. (*Figure 3* shows some of these templates).

Occasions often arise where the rules break down in generating an alignment. This is generally true where the novel sequence is very different from the template (e.g. in the generation of peptide receptor models from neurotransmitter templates). In these cases the user has the option of moving the sequences into a visual alignment on the screen with the mouse. A typical alignment procedure is shown in *Figure 4.* By having the profile of the novel sequence displayed simultaneously, the user can

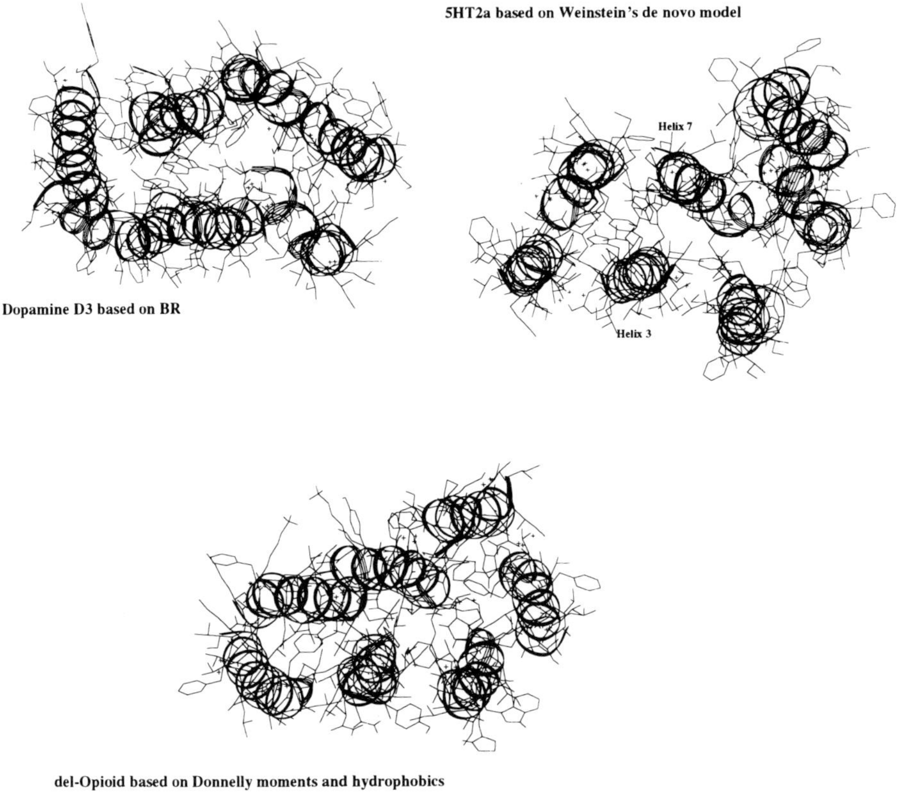

Figure 3. Three typical templates used in GPCR_Builder. BR, bacteriorhodopsin.

ensure that the chosen alignment falls within the predicted hydrophobic regions of the protein. Residues which are identical to the template are boxed to highlight them. At the same time the arrows above certain residues of the template sequence indicate the inward facing residues of that model. These can be used as additional aids in defining the alignment, particularly when used in combination with other tools which define the inward facing residues of a transmembrane helix (see, for example, the Helanal routine described later).

Once an alignment has been achieved, a starting three-dimensional coordinate file is constructed from the template using only atoms of the conserved residues and the backbone. Other residues have Cαs and Cβs only. Internal checks are made for inconsistencies such as glycines with Cβ or prolines with a backbone amide hydrogen, etc. All other CHARMm choices, such as minimization method, number of iterations, dihedral or atom constraints, etc., are chosen from the dialogue boxes. The job is then submitted automatically to a chosen machine on the network.

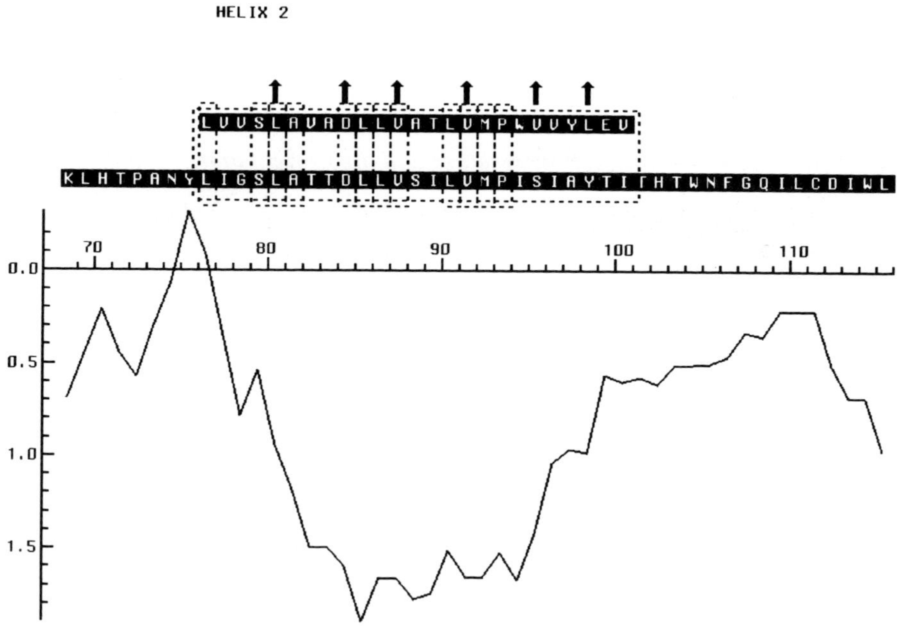

Figure 4. Aligning the template with the sequence. The hydropathic profile is shown below; residue matches are boxed. The inward-facing residues of the template are shown by the arrows on top.

Additional Tools in GPCR_Builder

In the previous section, the construction of a three-dimensional model of a GPCR from a predefined template was described. All the early templates used were based initially on bacteriorhodopsin. However, as has been previously noted, the homology of GPCRs with bacteriorhodopsin is very low and as a consequence, it was also our intention in this work to examine alternative models not based on bacteriorhodopsin. Such models might, for example, have different helix orientations or arrangements, or they might include contributions from their local environment. In the design of novel three-dimensional models, a number of problems can immediately be identified. Foremost amongst these are:

(1) Which residues are included in the helices?
(2) What determines the geometry and orientation of the helices?
(3) What can we do about the loops and the internal and external environment (e.g. internal solvation, the lipid bilayer, etc)?

The first two of these problems and the ways in which they have been dealt with in GPCR_Builder will be discussed together as they naturally follow on from each other. We have more recently started to look at the third topic and our results to date will be presented briefly.

Identification of transmembrane residues and their orientation

The most common methods used to determine the residues in the helices are profile analysis and identification of residue conservation. More recent advances in some of these techniques have been discussed in detail elsewhere in this book. Suffice it to say that the older statistically based methods for helix prediction such as Chou and Fasman (1974) do not work for membrane bound helices. Workers in this field have therefore relied on hydropathic analysis using profiles such as those of Kyte and Dolittle (1982) or Engelman *et al.* (1986). These methods simply assign a fixed value to each amino acid and average the hydropathicity along the sequence over a window of, say, 19 residues. All of these profile methods are readily entered into GPCR_Builder as external ASCII files and the user can set both the method and the window length.

The second method used, that is conservation of residues, is based on the concept that residues involved in the binding sites of the receptor (i.e. the transmembrane regions) will be more conserved than those in regions such as the interhelical loops which are not directly involved. (Note, however, that extracellular loop regions have been implicated in the binding of many peptide receptors). Furthermore, within the transmembrane regions themselves, it will be the inward-facing residues rather than those facing the lipid bilayer, which show the greatest degree of conservation. This conservation can therefore be used in two ways to help in the determination of GPCR structure. It can not only help define residues within the TM regions but it can also be used to define the orientation of the helix around its longitudinal axis.

Sequence conservation is best observed by visualization of multiple alignments using a suitable colour-coding scheme. This is especially useful when used in combination with profile analysis (*Figure 5*).

GPCR_Builder reads in multiple alignment files from the usual sources such as GCG, CLUSTAL and Seqsee. These are then displayed, with or without an average profile, as shown in *Figure 5*. They can readily be manipulated on the screen in real-time using the dials.

It was noted above that the inward-facing residues in the helix bundle should show a greater degree of conservation than those pointing into the lipid bilayer. The regular nature of an α-helix means therefore that this conservation should exhibit a periodicity. Using Fourier transform techniques, Donnelly *et al.* (1989) were able to demonstrate this periodicity and by defining a ‘moment’ of conservation, were able

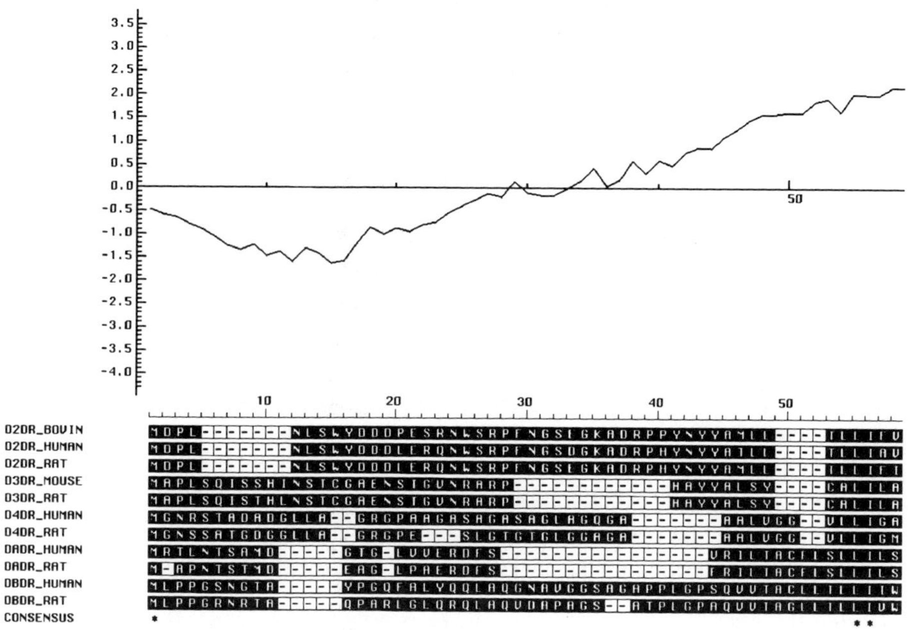

Figure 5. A CLUSTAL-V multiple alignment of some dopamine sequences with average profile.

to indicate the inward face of the helix. Another method which has been used in the prediction of helix orientation is the hydrophobic moment. In a transmembrane helix, this should point in the opposite direction to the conservation moment above (i.e. towards the lipid bilayer). However, at the membrane interface where the helices end and the loop regions begin, Donnelly and Cogdell (1993) predicted that the hydrophobic moment should reverse and point in the same direction as the conservation. The positions of reversal should therefore indicate the ends of the transmembrane regions.

Both the hydrophobic and the Donnelly conservation moments are calculated and can be displayed within GPCR_Builder. Based on the description of Baldwin (1994), an algorithm was developed (known internally as Helanal) which could quickly display the results of multiple sequence alignments as colour-coded helical wheels (*Figure 6*, see page 110). Included in the wheel was information about which residue positions were identical or conserved, the physicochemical type (i.e., aromatic, acidic, basic, etc.) of the residue at a given position and the 'Donnelly' and/or hydrophobic moments. An estimate of how buried a given helix was, could be made from the extent of the hydrophobic arc of the wheel. Thus, by simultaneously displaying the Helanal wheels for the seven TM regions and manipulating them so that the 'Donnelly moment' pointed inward and hydrophobic arcs remained exposed, a good two-dimensional representation of the GPCR could be obtained. These Helanal wheels are the basis for the construction of idealized three-dimensional helices which as will be seen later, are being used to study hydrophobic properties of the protein.

In our earliest work in the construction of three-dimensional models of GPCRs (Blaney, 1991), no account was taken of the perceived homology with the bacteriorhodopsin structure. Rather, we made use of the concept of hydrophobic energy as described by Eisenberg and McLahlan (1986). According to this description each atom in a protein has associated with it, an atomic solvation parameter (ASP). Thus, the contribution of that atom to the overall hydrophobic energy will be the product of its ASP and its accessible surface area. A program was developed (Blaney, 1991) which calculated the analytical surface area and its atomic components for the seven helical bundle and from these, calculated the overall hydrophobic energy using the ASPs of Eisenberg and McLahlan (1986). This energy was then maximized by alteration of the translational and rotational degrees of freedom of the individual helices, considered as single units. The problems associated with the method can be summarized as shown in *Figure 7* (see page 111).

During the maximization procedure, there is a natural tendency for the helices to move apart, hence increasing the overall accessible surface area. For this reason also, there was no driving force for the helices to tilt as this could result in a decrease in the surface area of the bundle. The end result was a GPCR model with fully parallel helices and an unrealistically large interior cavity when compared with bacteriorhodopsin. The method was therefore abandoned at that time in favour of models generated by alignment with bacteriorhodopsin.

Recent examination of the hydrophobic arcs in the Helanal algorithm suggested that there was considerable variation in the helices of GPCRs which could give rise to slightly different packing arrangements of the bundle. This is confirmed by the publication of a low-resolution electron diffraction map of bovine rhodopsin (Schertler *et al.*, 1993) which shows a slightly different helix arrangement to that of bacteriorhodopsin. In considering alternative models for determining interhelical tilting we have decided once again to make use of calculations of hydrophobic properties associated with the helix surface. Rather than compute an overall energy, we opted to use a hydrophobic potential energy associated with each point on the surface. This, it was hoped, would better differentiate between various areas of the helix. One should therefore, in theory, see a gradation in going from the internal face of the helix to the surface pointing to the lipid bilayer. It was also expected that one might expect a variation in the hydrophobic potential which would predict a tilt of that helix relative to the bilayer.

Within GPCR_Builder, using the information from the Helanal algorithm above, an idealized three-dimensional α-helix can be generated which includes the 'Donnelly moment' vector. To determine the hydrophobic potential, the analytical surface area of each atom within this helix is calculated using the Richmond algorithm (Richmond, 1984). The enthalpy and free energy associated with each atom was obtained using the atomic solvation parameters of Ooi *et al.* (1987) (*Table 1*). The Connolly solvent accessible surface is then calculated and at each point on the surface the potential P_i, is computed as

$$P_i = \varepsilon_i + \sum_{j \neq i}^{num\_atoms} \frac{\varepsilon_j}{r_{ij}}$$

where ε_i is the energy associated with atom *i*, calculated from the product of the atomic solvation parameter and the accessible surface area, and r_{ij} is the distance between point *i* and the atomic centre *j*. This choice of a 1/r dependence of the hydrophobic potential on atomic distance was made by analogy with electrostatic potential. There is no compelling reason however not to choose an alternative distance function. Several other groups have described other measures of hydrophobic potential. Fauchère *et al.* (1988) used a simple exponential distance term ($e^{-\alpha r}$) whereas Kellogg andAbraham (1992) initially used a ($1/r^2$) descriptor in their HINT program before reverting to an exponential term. Finally, the potential is colour-coded (blue for hydrophobic, red for hydrophilic) on to the meshed or solid surface for display (*Figure 8,* see page 111).

By interactively manipulating the whole set of seven colour-coded helices on the screen to visually maximize the blue hydrophobic surface on the exterior of the bundle and keep the three-dimensional 'Donnelly moment' pointing to the interior, one can generate new starting arrangements of the helices. This technique is still

Table 1. The atomic solvation parameters used in the calculation of hydrophobic potential (from Ooi *et al.*, 1987)

Chemical class	van der Waal's radius	g (kcal mol^{-1} Å^{-1})	–h (kcal mol^{-1} Å^{-1})
Aliphatic CH_3,CH_2,CH	2.0	0.008	0.026
Aromatic CH	1.75	–0.008	0.038
Hydroxyl OH	1.40	–0.172	0.238
Amide amine NH	1.55	–0.132	0.192
Carboxyl carbonyl C	1.55	0.427	–0.413
Carboxyl carbonyl O	1.40	–0.038	0.032
Sulphur	2.0	–0.021	

currently under evaluation, but has already produced some interesting new GPCR models which differ from those based on bacteriorhodopsin alignments.

Simulation of the protein environment

Another area which we have been investigating with GPCR_Builder is that of simulating the protein environment. This has involved a study both of the internal solvation in the helix bundle and of the exterior lipid bilayer/water medium. There are two main reasons why we may want to solvate the GPCR models.

(1) It is likely that the inside of the receptor, being quite polar, does contain some water molecules. These may even conceivably be involved in bridged hydrogen bonding between the receptor side chains and the ligand, which has frequently been found to occur with globular proteins. The inclusion of interior water should therefore make the model more realistic.

(2) One of our fundamental reasons for generating GPCR models is to use these in the *de novo* design of novel potentially therapeutic ligands. For these purposes a number of programs such as SPROUT, LUDI, HOOK, CAVEAT (Gillet *et al.*, 1994; Bohm, 1992; Molecular Simulations Inc.; and Lauri and Bartlett, 1994, respectively), etc. have emerged over the last few years. All of these programs depend to some extent on their ability to explore the available cavities around the active site of a protein and then generate potential structures which will fit into them. During the initial minimization of our GPCR models, the side chains were frequently observed to fill any cavities. Thus there was no available space within the active site to carry out ligand design studies. It was reasoned that the inclusion of interior water would go some way towards correcting this situation.

The method which is being used to place the interior waters is based on the generation of a completely presolvated helix bundle in which residues are gradually mutated from alanine to their proper sequence. Thus the initial helix bundle is constructed with only the glycine and proline residues intact. All other residues are set up as alanine. These are then gradually mutated to their correct residues starting with the polar residues, then the aromatics and finally the hydrophobic residues. Short minimization steps are carried out during this mutation and interior waters are removed only when clashes occur.

The main algorithmic problem in this approach is to define which waters are inside the receptor and which are outside. This has been achieved by geometrically defining the skewed surfaces between each helix and its two neighbours. Waters which are on the outside of these surfaces are then removed (*Figure 9*). The actual algorithm used for this is outlined below.

(1) The receptor helix bundle is placed in a 45 Å box of water. Outside waters lying beyond the maximum dimensions of the receptor are then removed.

(2) For each helix then:

(a) define the helix axis, translate its centre to the origin and align this vector with its extracellular face along the positive z axis.

(b) calculate the vector from the origin to the centre of the remaining six helices. Rotate to align this vector along the positive x axis. This results in four possible quadrant arrangements where the two neighbouring helices can lie.

(c) now define the skewed surface from the current helix to each of its two neighbours.

(d) for each water in turn, project its z value on to the surface and define the vector of the corresponding line on this surface.

(e) if the x value is less than that on the surface then it is considered outside and is removed.

This methodology has been carried out with a number of cationic neurotransmitter receptor models and in each case a total of 80–100 water molecules have ended up in the interior cavities. This is obviously too dense to approach the situation in

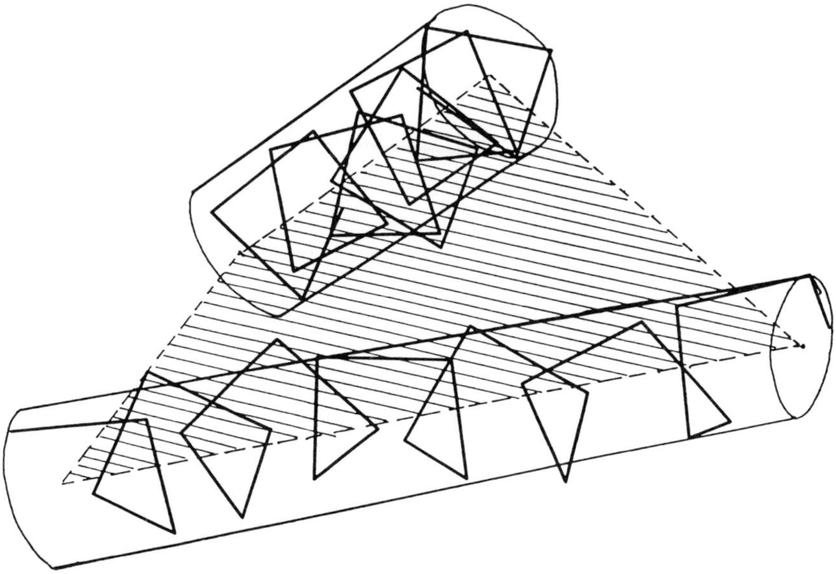

Figure 9. Two non-parallel helices with the skewed suface defined between their principal axes. Waters on one side of this will be defined as outside, depending on the orientation of this surface to the rest of the bundle.

nature. However, it has meant that the cavities have been well maintained and these models have been useful in application of *de novo* design tools.

Finally, in addition to the interior environment, we have also been looking at some simplified simulations of the exterior. Obviously the ideal situation would be to emerse the whole protein, complete with glycosylated loops, and with its interacting G protein, into a lipid bilayer. Layers of extracellular and cytoplasmic water would be added and the whole ensemble enclosed in periodic boundaries. Unfortunately, the huge number of atoms involved alone would make such a dynamics simulation impossible even on modern supercomputers. Furthermore, simulations in the order of nanoseconds would be necessary for this to give any meaningful results. To simplify this situation we have been investigating the use of langevan dynamics on a model system. This was generated with a single lipid bilayer shell. Positional restraints were applied to the terminal methyl groups of the hydrocarbon chains to prevent them from flying apart. A model was constructed of the 5-HT_{2c} receptor to which all the loops except the large third cytoplasmic were added. These loops were generated from a search of the Brookhaven database using standard methodology. The first 10 residues of both N- and C- termini and the ends of the third cytoplasmic loop were also included but no geometrical secondary structure was imposed on these. To simulate the aqueous environment on either side of the membrane, hemispherical water caps were added. The entire ensemble has been included as a template in GPCR_Builder but the loops of the actual target receptor are built externally. The most important aspect of this study was to mimic the forces from the surrounding bulk lipid and water. To do this, frictional coefficients, calculated using Stoke's law, were applied to the exposed lipid and outer water atoms. Finally a molecule of the agonist, lysergic acid diethylamide (LSD), was docked into the putative binding site in the receptor channel.

After minimization, the system was allowed to evolve for 510 ps at 300 K, using a time step of 1.5 fs. As usual, the SHAKE algorithm was used to constrain hydrogen bond lengths. During the simulation, the helices were observed to bend at proline and glycine residues, developing pronounced kinks after the simulation. Interestingly, water was found to diffuse down the receptor channel from the extracellular side of the receptor, forming a hydrogen-bonded network from the bulk solvent to the active site. No water however was seen to diffuse from the intracellular side of the receptor into the channel.

Unfortunately, the side chains of the helices exposed to the lipid did not exhibit the behaviour of a dampened system, as would have been expected if the receptor were in a lipid proper. It appears therefore that the frictional coefficients do not mimic the bulk lipid in the correct manner, and hence do not exert enough force on the receptor to dampen the side chain motion. While the early results do not appear to be overly encouraging, work continues in this area in an attempt to simulate the lipid forces more correctly.

4. Conclusion

The GPCR_Builder program has proved to be of great use both in the construction of receptor models based on existing templates and in the construction of novel helix bundle packings. Using these models it has been possible to rationalize the binding of both agonists and antagonists in a number of receptor systems and they are now being used in the design of novel potential therapeutic agents. As more ligand binding information becomes available our faith in these models is expected to increase, but we still eagerly await the results for the first definitive structure of a GPCR. The publication by Schertler *et al.* (1993) gives us some hope that this may not be in the too distant future.

Acknowledgements

We are grateful to Professor Harel Weinstein (Mount Sinai Hospital, New York) for giving us the coordinates of his 5-HT_{2a} model.

References

Baldwin JM. (1994) Structure and function of receptors coupled to G proteins. *Curr. Opin. Cell Biol.* **6**: 180–190.

Blaney FE. (1991) A unified model for serotinin-5HT2 antagonist receptor binding. Paper presented at the International Society of Quantum Biology and Pharmacology (ISQBP) presidential meeting, Stanford, California, USA.

Böhm HJ. (1992) The computer program LUDI: a new method for the de novo design of enzyme inhibitors. *J. Comp. Aided Mol. Design,* **6**: 61–78.

Brooks BR, Bruccoleri RE, Olafson BD, States DJ, Swaminathan S, Karplus M. (1983) CHARMM. A program for macromolecular energy, minimisation and dynamics calculations. *J. Comput. Chem.* **4**: 187–217.

Chou PY, Fasman GD. (1974) Prediction of protein conformation. *Biochemistry,* **13**: 222–245.

Clark T, Rauhut G, Chandrasekhar J. (1991) University of Erlangen.

Donnelly D, Cogdell RJ. (1993) Predicting the point at which transmembrane helices protrude from the bilayer: a model of the antenna complexes from photosynthetic bacteria. *Prot. Eng.* **6**: 629–635.

Donnelly D, Johnson MS, Blundell TL, Saunders J. (1989) An analysis of the periodicity of conserved residues in sequence alignments of G protein-coupled receptors – Implications for the 3 dimensional structure. *FEBS Lett.* **251**: 109–116.

Eisenberg D, McLahlan AD. (1986) Solvation energy in protein folding and binding. *Nature,* **319**: 199–203.

Engelman DM, Steitz TA, Goldman A. (1986) Identifying non-polar transbilayer helices in amino-acid sequences of membrane proteins. *Ann. Rev. Biophys. Chem.* **25**: 321–353.

Evardsen O, Sylte I, Dahl S. (1991) Molecular dynamics of dopamine at the D_2 receptor. *Proc. Natl Acad. Sci. USA,* **88**: 8111–8115.

Evardsen O, Sylte I, Dahl S. (1992) Molecular dynamics of serotonin and ritanserin interacting with the 5-HT_2 receptor. *Mol. Brain Res.* **14**: 166–178.

Fauchère J, Quarendon P, Kaetterer L. (1988) Estimating and representing hydrophobicity potential. *J. Mol. Graphics,* **6**: 203–206.

Gillet VJ, Newell W, Mata P, Myatt G, Sike S, Zsoldos Z, Johnson AP. (1994) Recent developments in the *de novo* design of molecules. *J. Chem. Inf. Comp. Sci.* **34**: 207–217.

Kellogg GE, Abraham DJ. (1992) KEY, LOCK, and LOCKSMITH: Complementary hydropathic map predictions of drug structure from a known receptor–receptor structure from known drugs. *J. Mol. Graphics*, **10**: 212–217.

Kyte J, Dolittle RF. (1982) A simple method for displaying the hydropathic character of a protein. *J. Mol. Biol.* **157**: 105–132.

Lauri G, Bartlett, PA. (1994) CAVEAT: a program to facilitate the design of organic molecules. *J. Comp. Aided Mol. Design*, **8**: 51–66.

Lewell XQ. (1992) A model of the adrenergic beta-2 receptor and bindig sites for agonist and antagonist. *Drug Design Discovery*, **9**: 29–48.

Livingstone CD, Strange PG, Naylor LH. (1992) Molecular modeling of D2-like dopamine receptors. *Biochem. J.* **287**: 277–282.

MaloneyHuss K, Lybrand TP. (1992) Three-dimensional structure for the β_2 adrenergic receptor protein based on computer modeling studies. *J. Mol. Biol.* **225**: 859–871.

Ooi T, Oobatake M, Nemethy G, Scheraga HA. (1987) Accessible surface areas as a measure of the thermodynamic parameters of hydration of peptides. *Proc. Natl Acad. Sci. USA*, **84**: 3086–3090.

Probst WC, Snyder LA, Schuster DI, Brosius J, Sealfon S. (1992) Sequence alignment of the G-protein coupled receptor superfamily. *DNA Cell Biol.* **11**: 1–20.

Richmond TJ. (1984) Solvent accessible surface area and excluded volume in proteins. Analytical equations for overlapping spheres and implications for the hydrophobic effect. *J. Mol. Biol.* **178**: 63–89.

Schertler GFX, Villa C, Henderson R. (1993) Projection structure of rhodopsin. *Nature*, **362**: 770–772.

Trumpp-Kallmeyer S, Hibert MF, Bruinvels A, Hoklack J. (1991) Three-dimensional models of neurotransmitter G-binding protein-coupled receptors. *Mol. Pharmacol.* **40**: 8–15.

Trump-Kallmeyer S, Hoklack J, Bruinvels A, Hibert MF. (1992) Modeling of G-protein-coupled receptors: application to dopamine, adrenaline, serotonin, acetylcholine and mammalian opsin receptors. *J. Med. Chem.* **35**: 3448–3462.

Vinter JG, Davis A, Saunders MR. (1987) Strategic approaches to drug design 1. An integrated software framework for molecular modelling. *J. Comp. Aided Mol. Design*, **1**: 31–51.

Zhang D, Weinstein H. (1993) Signal transduction by a 5-HT_2 receptor: a mechanistic hypothesis from molecular dynamics simulations of the three-dimensional model of the receptor complexed to ligands. *J. Med. Chem.* **36**: 934–938.

Chapter 11

Fluorescence studies of neurokinin receptors

André Chollet, Gerardo Turcatti, Karin Nemeth and Charles Bradshaw

1. Introduction

A large number of biologically active peptides mediate their cellular effects by integral membrane proteins belonging to the superfamily of G protein-coupled receptors (GPCRs) which are predicted to have seven transmembrane regions. These peptides range in size from tripeptides, such as formyl peptides or thyrotropin-releasing hormone, to much larger peptides, such as neuropeptide Y or chemokines. An understanding of the structural basis of ligand–receptor recognition would provide information on the molecular mechanisms of receptor activation and modulation of signal transduction. For example, the mammalian neurokinin (tachykinin) system represents a typical example of functional diversity within a group of closely related and biologically active peptides and their respective receptors (Nakanishi, 1991).

Getting a three-dimensional structure of a neurokinin receptor, or any seven-transmembrane receptor, represents a formidable task because of the difficulties in obtaining large amounts of protein and in producing crystals that diffract X-rays. In the absence of high resolution structure, molecular models have been created. These models are based on electron cryo-microscopy structural data on bacteriorhodopsin and rhodopsin. The relevance of models based on bacteriorhodopsin, a bacterial proton pump that does not share primary structure homology with GPCRs, is questionable, and structural data on rhodopsin is available only at low resolution. Consequently, there is a strong need to develop novel methods and approaches for the biophysical analysis of seven-transmembrane receptors. An X-ray structure would probably not provide information in real-time about the conformational changes that occur during the functional activation of the receptor. Fluorescence spectroscopy techniques, such as fluorescence energy transfer or fluorescence anisotropy, are sensitive methods that have the potential to give insight into dynamic structural changes during ligand binding and receptor activation, as well as to provide experimental data to help molecular modelling. In this chapter, we describe our fluorescence spectroscopic studies of the interaction of fluorescent peptide ligands with the neurokinin NK2 (neurokinin A, NKA, substance K) receptor to probe ligand binding domains and we discuss application of fluorescence techniques to probe the receptor folding and function.

2. Methods

Fluorescent peptides

The synthesis, purification, and chemical and pharmacological characterization of each fluorescent peptide analogue was carried out according to previously reported procedures (Bradshaw *et al.*, 1994; Ceszkowski and Chollet, 1992).

Cell culture

The CHO/A cell line expressing 700 000 NK2 receptors per cell was established and cultured as previously described (Turcatti *et al.*, 1993). CHO/T cells (45 000 NK2 receptors per cell) were cultured as reported earlier (Bradshaw *et al.*, 1994). For spectrofluorimetry and microscopy experiments with cells in suspension, cells were grown to 80% confluence and harvested in phosphate-buffered saline (PBS), pH 7.4, containing 1mM EDTA and washed with ice-cold PBS.

Flow cytometry

Experiments were performed as described previously (Bradshaw et al., 1994).

Confocal microscopy

Approximately 10^6 CHO/A cells were incubated in 100 μl PBS containing 1.4 nM fluorescent ligand, in the presence or absence of 20 μM GR94800, for 1 h at ambient temperature. After two washes with PBS, the cells were suspended in 100 μl PBS and analysed under a confocal microscope (Zeiss, magnification 720×; 1994) using an argon laser for excitation at 488 nm and recording emission at 510–525 nm.

Spectroscopy

UV-visible absorbance spectroscopy measurements were made using a Hewlett-Packard 8452A diode array spectrophotometer at ambient temperature. Steady-state fluorescence measurements were recorded using a Jasco FP-777 spectrofluorimeter with a 0.5 × 1.0 cm quartz cuvette of 500 μl sample volume, stirred continuously with a magnetic bar. Fluorescence measurements were made on cells in suspension at 20°C. The excitation wavelength was 475 nm and the emission was recorded either at 540 nm (7-nitrobenz-2-oxa-1,3-diazol-4-yl, NBD) or at 520 nm (fluorescein). The excitation and emission passing bands were 5 nm and 10 nm respectively.

Collisional quenching experiments

About 10^6 cells were incubated in PBS, pH 7.2, containing $MgCl_2$ (3 mM), bovine serum albumin (BSA) (0.2 mg ml^{-1}), in the presence of 20 nM fluorescent ligand at 20°C for 1 h. Cells were then washed twice with ice-cold PBS, pH 7.2, and resuspended in PBS for fluorimetry measurements. Fluorescence collisional quenching experiments were performed by adding increasing amounts of a 0.153 M KI solution in water to a final concentration of 50 mM in a cuvette containing 10^6 suspended cells in an initial volume of 500 µl. Changes in fluorescence due to the addition of quencher were corrected by subtracting the fluorescence measured in parallel samples in which quencher was added to cells in PBS or cells saturated with an excess (500 to 1000-fold) of non-fluorescent antagonist GR100679 (Smith *et al.*, 1993) before exposure to the fluorescent ligand. The quenching of the fluorescence emission at 540 nm was calculated using the Stern–Volmer equation (Lakowicz, 1983): $F_0/F = 1 + K_{SV}\,[\,I^-]$, where F_0/F is the ratio of fluorescence intensities in the absence and presence of iodide. The Stern–Volmer quench constant (K_{SV} as determined from the slope of F_0/F as a function of the iodide concentration $[\,I^-]$.

Real-time monitoring of ligand binding to NK2 receptor

About 2×10^6 CHO/A cells were suspended in 500 µl of PBS containing 3 mM $MgCl_2$ and 20 mg $ml^{-1}$of BSA at 20°C (about 2 nM [^{3}H]GR94800 binding sites). Fluorescence binding measurements were initiated by addition of 4 nM fluorescent antagonist DAB-N. The binding was monitored on the spectrofluorimeter at 540 nm. After completion of binding (constant value of fluorescence emission), a 1000-fold molar excess of the competitive antagonist GR100679 or GR 94800 was added and dissociation of DAB-N was monitored over time. Contribution due to free ligand was subtracted where necessary by comparison with samples preincubated with 2 µM GR100679. Kinetic parameters for association were obtained from the plot of ln $[B_{eq}/B_{eq} - B_t]$ versus time, where B_{eq} is the fraction bound at equilibrium and B_t is the fraction bound at time t during the association. Dissociation kinetic constants were obtained from the plot of ln B_t/B_0 versus time where B_t is ligand bound at time t and B_0 is ligand bound at time $t = 0$ (Limbird, 1986).

In vitro *expression of NK2 receptor*

A vector (pGEMNK2) for the efficient transcription of cDNA encoding the human ileum NK2 gene was constructed by engineering the commercially available pGEM2 vector (Promega Corp., Madison, WI, USA). Firstly, pGEM2 was cleaved with *Eco*RI and *Bam*HI and a synthetic 27-mer *Eco*RI-*Bam*HI linker containing a short 5′ non-translated region with low secondary structure forming potential (Zozulya *et al.*, 1990) and a *Nde*I site was inserted. The NK2 cDNA (Arkinstall *et al.*, 1994) was then subcloned, downstream of the SP6 polymerase initiation site, as a *Nde*I-*Bam*HI 1.26 kb fragment containing the ATG initiator at the *Nde*I site and a 62 bp3′ non-translated region upstream of the *Bam*HI site. Secondly, to improve transcription efficiency, a 27 bp *Eco*RI-*Ppu*MI restriction fragment encompassing the starting ATG codon was excised and replaced by a synthetic 27 bp *Eco*RI-*Ppu*MI fragment containing an eukaryotic consensus sequence for efficient initiation of translation at AUG (Kozak, 1987). The constructions were confirmed by restriction mapping and DNA sequencing. Run-off transcription *in vitro* was performed by SP6 polymerase on plasmid pGEMNK2 linearized by *Bam*HI as reported earlier (Gurevich *et al.*, 1991) to give a 1.3 kb NK2 mRNA transcript containing a short 5′ non-translated region with low secondary structure forming potential. The yield of mRNA was 300–400 μg from 10 μg plasmid. The mRNA was used directly in the *in vitro* translation assays: (a) in rabbit reticulocyte lysate: mRNA transcripts (2–3 μg) were incubated for 1 h at 25°C in a final volume of 10 μl comprising 6 μl rabbit reticulocyte lysate (Promega), 0.2 μl of a 1 mM mixture of all amino acid except methionine, 6 U RNAGuard (Pharmacia), and 0.5 μl [^{35}S]methionine (800 Ci mmol^{-1}, Amersham, final concentration 0.3 mCi ml^{-1}). The reaction was stopped on ice. The translation level was determined by measuring the incorporation of radioactivity into TCA-precipitated material. (b) In wheat germ extracts: mRNA transcripts (2–3 μg) were incubated for 1 h at 25°C in a final volume of 10 μl comprising 5 μl of wheat germ extract (Promega) containing 50 mM potassium acetate, 0.8 μl of a 1 mM mixture of all amino acids except methionine, 6 U RNAGuard (Pharmacia), and 0.5 μl [^{35}S]methionine (800 Cimmol^{-1}, Amersham). The reaction was stopped on ice. The level of translation was determined by measuring the incorporation of radioactivity into TCA precipitated material. For translations in the presence of microsomes, dog pancreatic microsomes (Promega) were used at 0.7 μl per 10 μl translation mixture. Oocyte microsomal membranes were prepared as described (Kobilka, 1990). The final membrane pellet was resuspended directly in reticulocyte lysate (50–100 μg in 10 μl of translation mixture).

3. Fluorescent ligands for the NK2 receptor

Synthesis of ligands

Several fluorescent probes for the NK2 receptor were designed, synthesized and pharmacologically characterized. Heptapeptide analogues of the highly potent and selective NK2 antagonist GR94800 were synthesized and derivatized with either fluorescein or NBD to give the fluorescent ligands DAB-N to LYS-BN shown in *Table 1* (Bradshaw *et al.*, 1994). Compounds DAB-N to LYS-BN differ by the length of the spacer arm connecting the fluorophore to the peptide moiety (see later). The fluorescent ligands DAB-N to LYS-BN retained high binding affinities for the NK2 receptor and were highly selective antagonists for NK2 over NK1 (substance P) and NK3 (neurokinin B) receptors in a functional activation assay measuring Ca^{2+} mobilization from intracellular stores (Bradshaw *et al.*, 1994).

Table 1. Structure of the fluorescent ligands

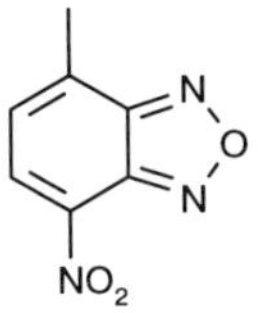

NBD

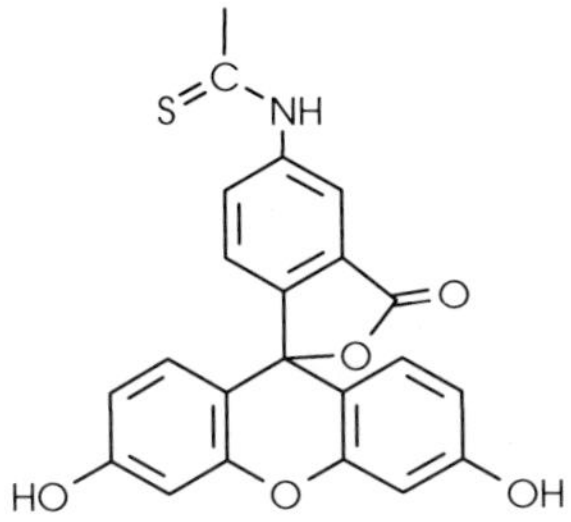

FLU

Compound	Structure*	d (Å)	pK_i†
GR94800	Bz-Ala-D-Trp-Phe-D-Pro-Pro-Nle-NH2	–	9.81 ± 0.07
DAB-N	Bz-**Dab(γ-NBD)**-Ala-D-Trp-Phe-R_1	4	8.87 ± 0.11
ORN-N	Bz-**Orn(**δ-NBD)-Ala-D-Trp-Phe-R_1	5	8.84 ± 0.07
LYS-N	Bz-**Lys(**ε-NBD)-Ala-D-Trp-Phe-R_1	6.5	8.83 ± 0.06
LYS-GN	Bz-**Lys(**ε-GlyNBD)-Ala-D-Trp-Phe-R	10	8.80 ± 0.03
LYS-AN	Bz-**Lys**(ε-ahNBD)-Ala-D-Trp-Phe-R_1	5	8.62 ± 0.17
LYS-BN	Bz-**Lys(**ε-bahNBD)-Ala-D-Trp-Phe-R_1	24	8.32 ± 0.02
NKA	His-Lys-Thr-Asp-Ser-Phe-Val-Gly-Leu-Met-NH_2	–	8.92 ± 0.04
N-NLE10-NKA[4–10]	**N-α (NBD)** Asp-Ser-Phe-Val-Gly-Leu-Nle-NH_2	–	8.08 ± 0.09
F-NLE10-NKA[4–10]	**N-α (FLU)** Asp-Ser-Phe-Val-Gly-Leu-Nle-NH_2	–	6.51 ± 0.20
N-NKA	**N-α (NBD)** His-Lys-Thr-Asp-Ser-Phe-Val-R_3	–	8.23 ± 0.01
F-NKA	**N-α (FLU)** His-Lys-Thr-Asp-Ser-Phe-Val-R_3	–	8.80 ± 0.15
DAP-7N	Ac-Asp-Ser-Phe-**Dap(β-NBD)**-R_2	–	5.71 ± 0.05

*R_1 = D-Pro-Pro-Nle-NH_2; R_2 = Gly-Leu-Nle-NH_2; R_3 = Gly-Leu-Met-NH_2; d = linearized distance between the NBD group and the Cα of amino acid at position 1; Bz = benzoyl; Nle = norleucine; Dab = 2,4-diaminobutyric acid; Orn = ornithine; Dap = 2,3-diaminopropionic acid; Ac = acetyl; ah = 6-aminohexanoyl; bah = bis(6-aminohexanoyl).
†Average value of 2–8 independent experiments.

Fluorescent decapeptide agonists N-NKA and F-NKA were synthesized by selective labelling of NKA at a single site on the N-α terminal group (Ceszkowski and Chollet, 1992). The fluorescent heptapeptides N-NLE10-NKA[4–10] and F-NLE^{10}-NKA[4–10] were made by the same procedure from the heptapeptide Nle^{10}-NKA[4-10] (Rovero *et al.*, 1989). The ligands N-NLE10-NKA[4-10], N-NKA and F-NKA retained a relatively high affinity for the NK2 receptor compared with the natural ligand NKA, whereas the ligand F-NLE10-NKA[4–10] was considerably less potent. Ligands N-NLE10-NKA[4–10] to F-NKA all activate the NK2 receptor in transfected CHO cells in a functional Ca^{2+} mobilization assay and this activity can be blocked by the selective NK2 antagonist GR94800. This indicates that the tagging of NKA or of Nle^{10} NKA[4–10] has not impaired their agonistic activity. In the ligand DAP-7N, (L)-diaminopropionic acid (Dap) was designed to replace Val^7 in Nlc^{10} NKA[4–10], and the newly introduced side-chain amino group was derivatized with NBD. The affinity of DAP-7N for NK2 was significantly reduced, suggesting that unfavourable steric or electronic interactions occur between NBD and residues in the binding pocket.

Fluorescence properties

Fluorescein and NBD were selected as fluorophores because of their high absorbance and fluorescence properties. The fluorescence of these groups is sensitive to the microenvironment and this feature can be used as an indication of the polarity of the environment. Fluorescein fluorescence is sensitive to pH, whereas NBD fluorescence is quenched by water and is therefore of higher intensity in a hydrophobic environment (Lancet and Pecht, 1977). These features can be exploited to physically characterize the ligand binding pockets on receptors (see below; Crowley *et al.*, 1993; Fay *et al.*, 1993). Indeed, the fluorescence emission intensity of DAB-N was 30-fold higher in non-polar solvents (e.g. chloroform) than in phosphate-buffered saline.

Visualization of NK2 receptors

The ability of the fluorescent ligands listed in *Table 1* to recognize NK2 receptors was analysed by flow cytometry and confocal microscopy on stably transfected CHO cells expressing the human ileal NK2 receptor at approximately 45 000 receptors/cell (CHO/T cell line) or 700 000 receptors/cell (CHO/A cell line) (Arkinstall *et al.*, 1994; Turcatti *et al.*, 1993). Representative fluorescence histograms for binding of a fluorescent NBD antagonist (DAB-N) or fluoresceinylated agonist (F-NKA) are shown in *Figure 1*. The specific fluorescence was measured as the mean fluorescence intensity of total cells with a lower limit of detection at approximately nanomolar concentrations of fluorescent ligand. CHO/A cells were analysed by three-dimensional confocal fluorescence microscopy (Brakenhoff *et al.*, 1989). *Figure 2* (see page 112) shows a fluorescence image of CHO/A cells labelled with the fluorescent antagonist DAB-N. In these overexpressing cells, the receptor is predominantly distributed on the cell surface and in a well-localized intracellular compartment, probably the Golgi apparatus, an organelle required for protein maturation before export. These results indicate that the fluorescent probes described above should be valuable tools for the localization and characterization of NK2 receptors in tissues.

4. Ligand–receptor interactions

Kinetics of ligand binding and dissociation

We monitored the binding of the NBD-containing antagonist ligand DAB-N to the NK2 receptor in real-time, and its subsequent dissociation after addition of an

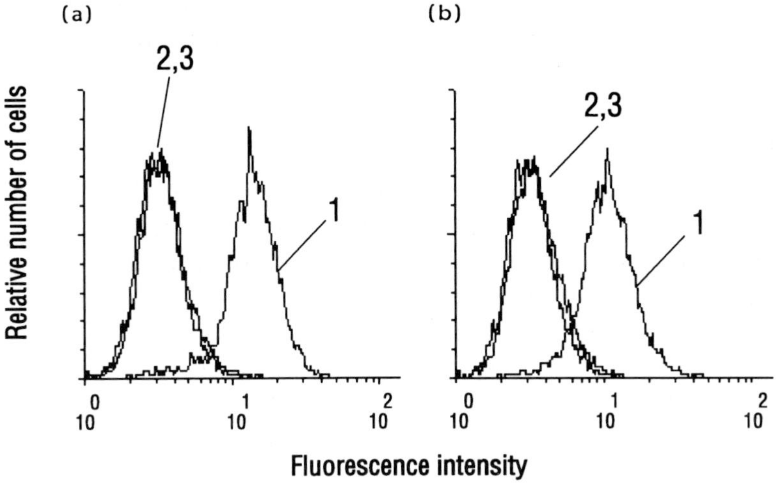

Figure 1. Flow cytometry analysis of NK2 receptors transfected in CHO cells using (a) the fluorescent antagonist DAB-N or (b) the fluorescent agonist F-NKA. 1 = 20 nM fluorescent ligand, 2 = 20 nM fluorescent ligand and 20 μM GR100679, and 3 = no ligand (autofluorescence of the cells). Excitation at 488 nm, emission at 518 nm.

excess of competitive antagonist. *Figure 3* shows the fluorescence at 540 nm as a function of time when DAB-N (4 nM final concentration) was added to suspended CHO/A cells in the absence or presence of excess NK2 antagonist GR94800. There was a rapid increase in fluorescence due to specific binding to NK2 receptors and equilibrium was reached after about 5–6 minutes. No specific binding was observed in the presence of antagonist. The kinetic parameters are listed in *Table 2*. The observed kinetics of association (k_{obs}) was calculated from the rate equation $k_{obs} = (\ln [B_{eq}/ B_{eq}-B_t])/t$. A linear plot was obtained ($r = 0.97$) from which a value of $k_{obs} = 0.53 \pm 0.1$ min^{-1} was calculated. The half-time for association was then calculated to be 1.4 min. These values were similar to the kinetics of [^{3}H]GR94800 binding to CHO/NK2 cells. The binding was reversible. Addition of an excess of the competitive antagonist GR100679 caused a slow decrease in affinity explained by dissociation of DAB-N from the NK2 receptor. The linear transformation of dissociation data using the equation $(\ln [B_t/B_0])/t$ fits best a biphasic curve with k_{diss} = 0.022 min^{-1} and 0.006 min^{-1}. This made a determination of the kinetic K_D inappropriate. The biphasic nature of the dissociation curve had been reported before for the tachykinins (Guard *et al.*, 1990). However, this does not indicate binding to two different sites since saturation binding analysis with the heptapeptide [^{3}H]GR94800 has revealed only one site (Turcatti *et al.*, 1993). Due to the lack of any change in fluorescence upon receptor binding (see below), it was not possible to determine the kinetics of agonist binding and dissociation by this method.

(a)

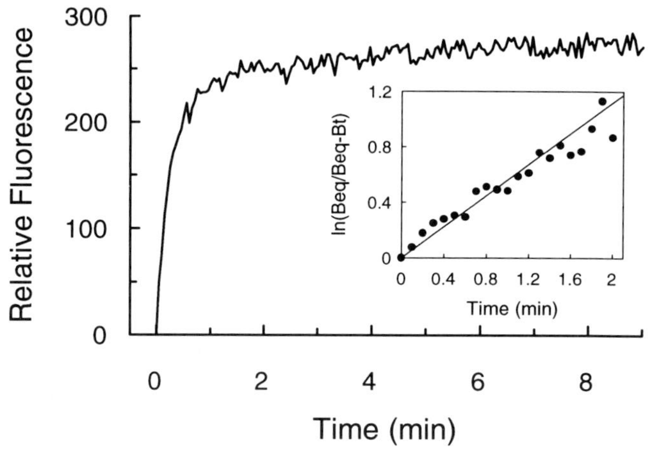

(b)

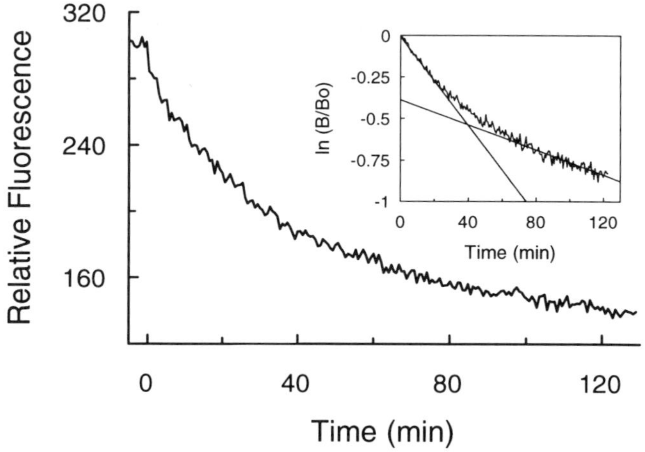

Figure 3. Real-time monitoring of (a) DAB-N (4 nM) binding and (b) dissociation from NK2 receptors on CHO cells. Inserts show the linear transformation of the data. The dissociation was initiated by addition of 40 µM GR94800. Inserts show the linear transformation of the data.

Investigation of ligand binding domains

Binding of fluorescent NBD ligands to the NK2 receptor was measured using a suspension of NK2-transfected CHO/A cells expressing about 700 000 receptors per cell. *Figure 4* shows a typical fluorescence emission spectrum for DAB-N. The quantum yield of NBD fluorescence emission was increased about five-fold when

Table 2. Kinetics of ligand binding to NK2 receptors on CHO cells

	Ligand association*	Ligand dissociation*	
Ligand	k_{obs} (min^{-1})	$k_{diss}1$ (min^{-1})	$k_{diss}2$ (min^{-1})
[^{3}H]GR94800	0.37 ± 0.08	0.036 ± 0.005	0.005 ± 0.003
DAB-N	0.53 ± 0.10	0.022 ± 0.006	0.006 ± 0.002

*Average values of 3–4 independent experiments.

the antagonist ligand DAB-N bound to the NK2 receptor on CHO/A cells, indicating that the fluorophore in the receptor-bound state was in a less polar environment than in the unbound state. In contrast, there was no increase in quantum yield of fluorescence emission for NBD-labelled agonists (N-NLE10-NKA[4–10], N-NKA) when they were allowed to bind to NK2 receptor, suggesting that the polarity of the binding pocket for the N-terminal portion of peptide agonists was similar to that of the aqueous solution. The specific fluorescence of fluorescein-labelled agonists F-NLE10-NKA[4–10] and F-NKA bound to NK2 was much higher than for N-NLE10-NKA[4–10] and N-NKA, even though F-NLE10-NKA[4–10] and F-NKA have lower binding affinities. Similar observations were made in flow cytometry experiments (data not shown). This indicates that the weak fluorescence signal observed for bound NBD agonists is principally due to the polarity of the environment.

To further define the binding pockets for agonist and antagonist peptides we carried out collisional quenching experiments. The fluorescence of all NBD-labelled ligands free in solution was quenched by the addition of increasing amounts of

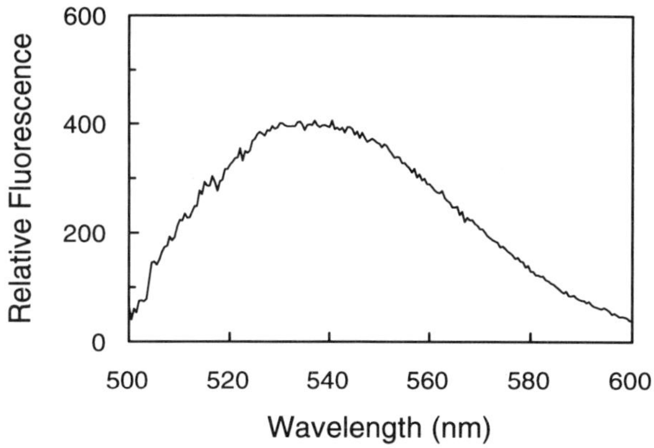

Figure 4. Spectrofluorimetric analysis of binding of NBD-labelled ligand DAB-N to NK2 receptors on CHO cells. Excitation at 475 nm. Specific fluorescence, obtained by subtracting fluorescence in presence of an excess of competing ligand from total fluorescence, is shown.

iodide to the solution. When the ionic strength of the solution was maintained at 0.153 M, a linear Stern-Volmer plot was obtained. This indicated a collisional quenching phenomenon. The Stern-Volmer constant, determined from the slope of F_0/F as a function of iodide concentration, is a measure of the degree of exposure of the NBD group to the solvent. Thus, solvent accessibility of the fluorescent ligands bound to the NK2 receptor was evalutated by measurement of NBD fluorescence quenching by KI. NK2 receptors on suspended CHO/A cells were saturated with fluorescent ligand, the excess of ligand was removed and iodide quenching was measured. F*igure 5* shows the Stern-Volmer plots obtained for various NBD antagonists and *Table 3* summarizes the experimental Stern-Volmer constants. Antagonist ligands DAB-N and ORN-N were not quenched at all indicating that the NBD group was not accessible to the solvent. In contrast, LYS-GN, LYS-AN and LYS-BN were quenched by iodide with Stern-Volmer constants comparable to those of ligands in free solution, indicating that the NBD group on these ligands was fully exposed to the solvent. LYS-N ligand was only partly quenched by iodide. *Figure 6* shows the relationship between quenching and the linearized length (*d*) of the spacer between the NBD group and the heptapeptide backbone in antagonist ligands. Ligands with $d < 5$ Å were inaccessible to quenching by iodide whereas ligands with $d > 10$ Å were efficiently quenched. Taken together these results indicate that the binding pocket for heptapeptide antagonists is shielded from the aqueous phase by between 5 and 10 Å.

Because of the weak specific fluorescence signal for N-NLE10-NKA[4–10] or N-NKA in the receptor-bound state (see above), collisional quenching experiments for agonists were performed with fluorescein-labelled F-NLE10-NKA[4–10] and F-NKA. As shown in *Table 3*, the fluoresceinyl group was quenched by iodide with the same efficiency in the receptor-bound state or free in solution, indicating that

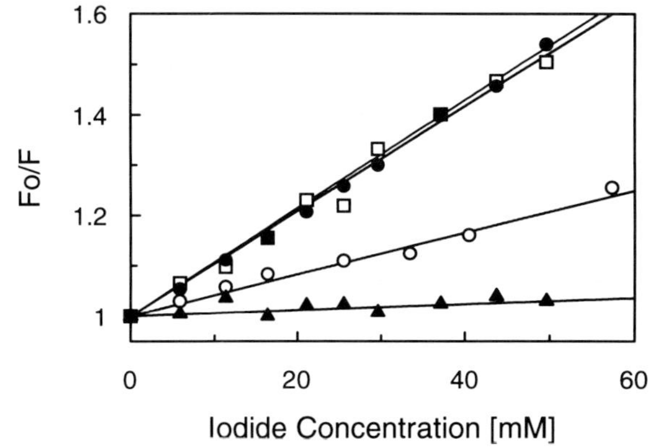

Figure 5. Stern–Volmer plots of quenching of NBD-labelled NK2 antagonists by KI. ● = LYS-N free in solution, ▲ = ORN-N, O = LYS-N, □ = LYS-BN.

Table 3. Stern–Volmer constants (K_{SV}) of quenching of fluorescent NK2 ligands by iodide*

Ligand	K_{sv} free (M^{-1})	K_{sv} bound (M^{-1})
DAB-N	10.9 ± 0.4	0.5 ± 0.3
ORN-N	10.8 ± 0.4	0.7 ± 0.3
LYS-N	10.7 ± 0.3	4.6 ± 0.5
LYS-GN	10.7 ± 0.5	9.8 ± 1.0
LYS-AN	10.8 ± 0.5	7.6 ± 0.8
LYS-BN	10.7 ± 0.3	9.8 ± 1.8
F-NLE10-NKA[4-10]	4.3 ± 0.5	4.5 ± 0.6
F-NKA	4.5 ± 0.5	4.2 ± 0.7

*K_{SV} values are average values of 3–8 independent experiments.

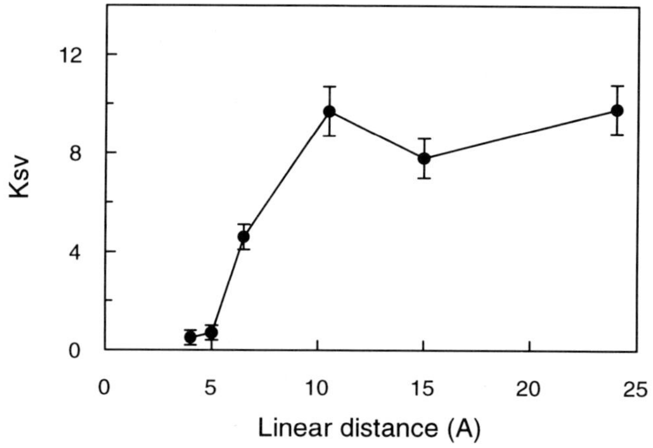

Figure 6. Efficiency of quenching of NBD-labelled NK2 antagonists by iodide as a function of the spacer length.

the N-terminal part of peptide agonists bound to the NK2 receptor is fully accessible to the aqueous solvent. To probe the environment of the C-terminal moiety of the peptide agonist, we designed and prepared the fluorescent ligand DAP-7N. Unfortunately, the affinity of DAP-7N for the NK2 receptor was too low for fluorescence measurements.

In summary, agonist and antagonist heptapeptide recognize binding sites on the NK2 receptor that are spatially non-identical and that differ in the nature of the binding determinants.

Implication for ligand–receptor recognition

Because of their size (10–11 amino acids), neurokinin peptides probably use several binding determinants spread over a relatively large area of their receptors. To understand better the molecular basis of these interactions, several groups have recently investigated the peptide binding site in neurokinin receptors using site-directed mutagenesis, chimeric receptors or N-terminally truncated receptors (reviewed by Fong and Strader, 1994). These studies indicated that multiple receptor domains are involved including the N-terminal segment, the extracellular loops and possibly the transmembrane regions. Little is known, however, about the relative importance of these residues. There is also evidence that different neurokinins may not use equivalent residues for binding to their respective receptors. Binding pockets for non-peptide NK1 and NK2 antagonists are distinct but overlap the SP and NKA binding epitopes (Gether *et al.*, 1993). There is virtually no information available regarding peptide antagonist recognition by tachykinin receptors.

Our spectrofluorimetric data of heptapeptide ligand binding to the NK2 receptor described above demonstrate different modes of recognition for antagonists and agonists. Antagonists bind into a hydrophobic pocket shielded from the solvent and located at a depth of 5–10 Å in the membrane. In contrast, agonists are accessible to the solvent when bound to NK2. These results support a model involving the extracellular loops of the receptor as the major binding determinant for agonist. For antagonists, it suggests that the binding site is located in the pore formed by the seven helical transmembrane domains which are especially rich in hydrophobic residues. We cannot exclude, however, that the binding pocket may simply be shielded by the extracellular loops connecting the helices. This model may operate in many peptide receptors and is supported by recent publications on formyl peptides receptors (Perez *et al.*, 1993), NK1 (Fong and Strader, 1994), thrombin receptors (Gerszten *et al.*, 1994) and IL-8 receptors (Leong *et al.*, 1994).

5. Probing receptor function by fluorescence

Fluorescence spectroscopy tools

As shown above, singly labelled fluorescent ligands are useful in defining receptor binding pockets. However, to investigate the receptor folding and function, one must be able to introduce extrinsic fluorescent markers at selected sites into the receptor protein itself. Fluorescence resonance energy transfer between a pair of donor and acceptor fluorophores located either in the receptor, in the ligand or even in the G protein will provide intra- or intermolecular distance measurements between the two fluorescent groups and make possible the monitoring of structural changes induced by ligand–receptor interactions (*Figure 7*). The use of collisional quenching techniques (see above) will provide information on solvent–membrane boundaries, on solvent-exposed loop regions and on the orientation of the membrane-spanning helices. Time-resolved fluorescence anisotropy can also be used to monitor conformational changes.

Fluorescence labelling of seven-transmembrane receptors

Introducing fluorescent groups at specific sites into seven-transmembrane proteins represents a major challenge. The classical approach of introducing reporter groups into proteins is by chemical modification of reactive cysteinyl or lysyl amino acid side chains. The method usually suffers from a lack of selectivity resulting in the uncontrolled introduction of label at multiple sites. Moreover, in the case of most seven-transmembrane receptors, it is seriously limited by the difficulties in production, solubilization and purification of active receptors. One exception has been the labelling of naturally abundant rhodopsin with nitroxide spin labels (Resek *et al.*, 1993).

One much more elegant approach is the biosynthetic incorporation of unnatural amino acids at specific sites into proteins using cell-free translation systems (Benner, 1994). In this recently developed methodology, a nonsense amber codon is created in the coding sequence, at the site selected for unnatural mutagenesis, and is suppressed during *in vitro* translation by using an amber suppressor tRNA, misacylated with the desired unnatural amino acid, to yield the mutant protein (Bain *et al.*, 1991; Noren e*t al.*, 1989). To date a variety of unnatural amino acids, including fluorescent probes (Cornish *et al.*, 1994), have been introduced into proteins using this methodology. This technique, however, has been limited to the synthesis of small soluble proteins (e.g. T4 lysozyme, p21ras) in *in vitro* systems like *E. coli* S30 extracts or rabbit reticulocyte lysates or wheat-germ extracts.

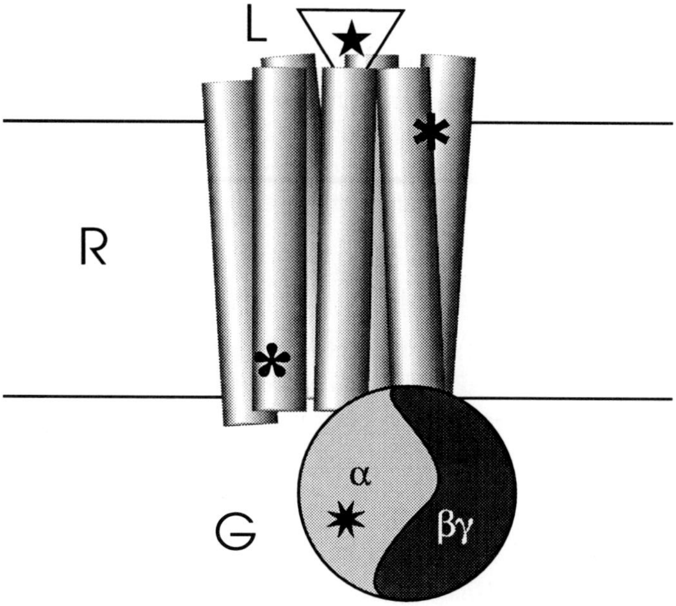

Figure 7. Schematic of fluorescence energy transfer study of receptor function. L = ligand, R = receptor, G = G protein. Stars represent fluorescent labels.

With the ultimate aim of using this methodology to produce nonnatural mutants of NK2, we expressed the NK2 receptor *in vitro* in two different mRNA-dependent cell-free translation systems, rabbit reticulocyte lysate and wheat-germ extracts. The translation products were labelled with [^{35}S]methionine and showed one major protein of apparent size 39 kDa on sizing gel electrophoresis. This size was comparable to glycosidase-treated NK2 receptor expressed in CHO cells (Turcatti *et al.*, 1993). When the translation was carried out in the presence of phosphatidylcholine liposomes (Zozulya *et al.*, 1990) most of the receptor was incorporated into the liposomes but was not functionally active as assessed by [^{3}H]GR94800 antagonist binding assays. This lack of binding activity suggests that missing membrane or cytosolic factors, potentially important for folding, translocation or post-translational modification, are necessary to confer functional activity on the NK2 receptors. Therefore, the receptor was expressed in reticulocyte lysates in the presence of either dog pancreatic microsomal membranes or microsomes prepared from *Xenopus laevis* oocytes (Kobilka, 1990). The NK2 receptor was found in microsomal membrane fractions and showed up as a diffuse band at 40–42 kDa on a sizing gel, indicating some post-translational modification. However, the NK2 receptor was not active in the ligand binding assay, thereby precluding us from exploring unnatural suppression mutagenesis techniques.

In vitro expression of functional seven-transmembrane receptors has been reported previously: bovine rhodopsin was expressed in wheat germ extracts in the presence of liposomes (Zozulya *et al.*, 1990) and β_2-adrenoceptor was expressed in rabbit reticulocyte lysate containing microsomes (Kobilka, 1990) . However, in both cases, the amount of successful expression depended on the batches of material and was not reliably reproducible. Secondly, only a small fraction of the expressed protein was found to be in an active state. This is a serious caveat if this approach is to be used to make fluorescent mutants because it would require tedious purification of the active receptor before any fluorescence analysis.

6. Conclusions and future directions

Understanding the structure and mechanism of action of GPCRs will require the integration of genetic, biochemical and biophysical approaches. In this chapter we describe methodology based on fluorescence spectroscopic techniques that gives insights into ligand–receptor recognition and that has the potential to probe conformational changes in receptor structure. Fluorescent groups on peptide ligands were used to characterize the binding site on the neurokinin NK2 receptor. By using environment-sensitive fluorescent probes and collisional quenching experiments, we have demonstrated that the ligand binding site for peptide antagonist on the NK2 receptor is hydrophobic and inaccessible to the solvent. Furthermore, by varying the length of the spacer arm tethering the NBD group to the peptide ligand we were able to measure the size and depth of the binding pocket. In contrast, peptide agonists bind to the NK2 receptor at a hydrophilic site that is accessible to the solvent. The methodology described here should prove generally useful for the study of ligand–receptor interactions in many different systems.

Functional activation of receptor involves dynamic changes in the structure. The introduction of fluorescence labels at specific sites in the receptor and study of the interaction of these labels with fluorescent groups on ligands or G protein will allow us to probe conformational changes and understand receptor function. With this goal in mind, we are progressing on the *in vivo* expression, in a cell system, of functionally active NK2 receptor containing an unnatural fluorescent amino acid analogue that has been introduced at a selected site by nonsense codon suppression technology.

Acknowledgements

We thank Dr Jean-Pierre Aubry for help with confocal microscopy and Guidon Ayala for the synthesis of oligonucleotides. We thank Drs Mike Edgerton and Horst Vogel for helpful discussion. We are grateful to Dr Jonathan Knowles for stimulating discussions and support throughout this work.

References

Arkinstall S, Emery I, Church D, Chollet A, Kawashima, E. (1994) Calcium influx and protein kinase Cα activation mediate arachidonic acid mobilization by the human NK-2 receptor expressed in Chinese Hamster Ovary cells. *FEBS Lett.* **338**: 75–80.

Bain JD, Diala ES, Glabe CG, Wacker DA, Lyttle MH, Dix TA, Chamberlin AR. (1991) Site-specific incorporation of nonnatural residues during *in vitro* protein biosynthesis with semisynthetic aminoacyl-tRNAs. *Biochemistry*, **30**: 5411–5421.

Benner SA. (1994) Expanding the genetic lexicon: incorporating non-standard amino acids into proteins by ribosome-based synthesis. *Trends Biotech.* **12**: 158–163.

Bradshaw C, Ceszkowski K, Turcatti G, Beresford I, Chollet A. (1994) Synthesis and characterization of selective fluorescent ligands for the neurokinin NK2 receptor. *J. Med. Chem.* **37**: 1991–1995.

Brakenhoff GJ, van Spronsen EA, van der Voort, HTM, Nanninga, N. (1989) Three-dimensional confocal fluorescence microscopy. In: *Methods in Cell Biology*, vol. 30 (Taylor DL, Wang YL, eds). San Diego: Academic Press, pp. 377–398.

Ceszkowski K, Chollet A. (1992) Synthesis of fluoresceinyl-neurokinin-A, a biologically active probe for NK2 receptors. *Bioorg. Med. Chem. Lett.* **6**: 609–612.

Cornish VW, Benson DR, Altenbach CA, Hideg K, Hubbell WL, Schultz PG. (1994) Site-specific incorporation of biophysical probes into proteins. *Proc. Natl Acad. Sci. USA*, **91**: 2910–2914.

Crowley KS, Reinhart GD, Johnson AE. (1993) The signal sequence moves through a ribosomal tunnel into a noncytoplasmic aqueous environment at the ER membrane early in translocation. *Cell*, **73**: 1101–1015.

Fay SP, Domalewski MD, Sklar LA. (1993) Evidence for protonation in the human neutrophil formyl peptide receptor binding pocket. *Biochemistry*, **32**: 1627-1631.

Fong TM, Strader CD. (1994) Functional mapping of the ligand binding sites of G protein-coupled receptors. *Med. Res. Rev.* **14**: 387–399.

Gerszten RE, Chen J, Ishii M, Ishii K, Wang L, Nanavicz T, Turck CW, Vu TKH, Coughlin SR. (1994) Specificity of the thrombin receptor for agonist peptide is defined by its extracellular surface. *Nature*, **368**: 648–651.

Gether U, Yokota Y, Emonds-Alt X, Breliere JC, Lowe JA, Snider RM, Nakanishi S, Schwartz TW. (1993) Two nonpeptide tachykinin antagonists act through epitopes on corresponding segments of the NK1 and NK2 receptors. *Proc. Natl Acad. Sci. USA*, **90**: 6194–6198.

Guard S, Watson SP, Maggio JE, Phon-Too H, Watling KJ. (1990) Pharmacological analysis of [^{3}H]-senktide binding to NK-3 tachykinin receptors in guinea-pig ileum longitudinal muscle myenteric plexus and cortex membranes. *Br. J. Pharmacol.* **103**: 1463–1469.

Gurevich VV, Pokrovskaya ID, Obukhova TA, Zozulya SA. (1991) Preparative in vitro mRNA synthesis using SP6 and T7 polymerases. *Anal. Biochem.* **195**: 207–213.

Kobilka BK. (1990) The role of cytosolic and membrane factors in processing of the human β-2 adrenergic receptor following translocation and glycosylation in a cell-free system. *J. Biol. Chem.* **265**: 7610–7618.

Kozak M. (1987) At least six nucleotides preceding the AUG initiator codon enhance translation in mammalian cells. *J. Mol. Biol.* **196**: 947–950.

Lakowicz JR. (1983) In: *Principles of Fluorescence Spectroscopy*. New York: Plenum Press, p. 131.

Lancet D, Pecht I. (1977) Spectroscopic and immunochemical studies with nitrobenzoxadiazolealanine, a fluorescent dinitrophenyl analogue. *Biochemistry*, **16**: 5150–5157.

Leong SR, Kabakoff RC, Hébert CA. (1994) Complete mutagenesis of the extracellular domain of interleukin-8 (IL-8) type A receptor identifies charged residues mediating IL-8 binding and signal transduction. *J. Biol. Chem.* **269**: 19343–19348.

Limbird LE. (1986) In: *Cell Surface Receptors: A Short Course on Theory and Methods*. Boston: Martinus Nijhoff Publishing, pp. 86–96.

Nakanishi S. (1991) Mammalian tachykinin receptors. *Annu. Rev. Neurosci.* **14**: 123–126.

Noren CJ, Anthony-Cahill, SJ, Griffith MC, Schultz PG. (1989) A general method for site-specific incorporation of unnatural amino acids into proteins. *Science*, **244**: 182–188.

Perez HD, Holmes R, Vilander LR, Adams RR, Manzana W, Jolley D, Andrews WH. (1993) Formyl peptide receptor chimera define domains involved in ligand binding. *J. Biol. Chem.* **268**: 2292–2295.

Resek JF, Farahbakhah ZT, Hubbell WL, Khorana HG.(1993) Formation of the meta II photointermediate is accompanied by conformational changes in the cytoplasmic surface of rhodopsin. *Biochemistry*, **32**: 12025–12032.

Rovero P, Pestellini V, Rhaleb NE, Dion S, Rouissi N, Tousignant C, Telemaque S, Drapeau G, Regoli D. (1989) Structure-activity studies of neurokinin A. *Neuropeptides*, **13**: 263–270.

Smith PW, McElroy AB, Pritchard JM, Deal MJ, Ewan GB, Hagan RM, Ireland SJ, Ball D, Beresford I, Sheldrick R, Jordan CC, Ward P. (1993) Low molecular weight neurokinin NK2 antagonists. *Bioorg. Med. Chem. Lett.* **3**: 931–936.

Turcatti G, Ceszkowski K, Chollet A. (1993) Biochemical characterization and solubilization of human NK2 receptor expressed in chinese hamster ovary cells. *J. Receptor Res.* **13**: 639–652.

Zozulya SA, Gurevich VV, Zvyaga TA, Shirokova EP, Dumler IL, Garnovskaya MN, Natochin MY, Shmukler BE, Badalov PR. (1990) Functional expression *in vitro* of bovine visual rhodopsin. *Protein Eng.* **3**: 453–458.

Chapter 12

A structural model for the 16 kDa membrane sector of the vacuolar H⁺-ATPase*

John B.C. Findlay, Malcolm E. Finbow, Philip C. Jones, Yong-In Kim and Michael A. Harrison

1. Introduction

The regulated movement of ions and other solutes across biological membranes is of fundamental importance to the living cell. Experimental strategies designed to provide an appreciation of the structure–function relationships of the proteins which facilitate these transport processes are, therefore, the subject of intense activity. Unfortunately, the membrane-bound proteins performing transport functions are usually very hydrophobic and frequently found at relatively low abundance in their native cell, making it difficult to obtain information by classical biophysical methods, at least in the short-term. The generation of structural models for membrane proteins using a variety of structure prediction approaches can provide a good approximation of protein structure. However, these models will remain insecure if unsupported by firm structural data..

Advances in molecular biology techniques in recent years have made it possible for relatively scarce membrane proteins to be cloned and expressed in heterologous systems. Techniques in site-directed mutagenesis have allowed structure–function studies to proceed without the requirement for protein isolated from the native system. By combining these molecular biology approaches with techniques in protein chemistry, it becomes possible to develop and test molecular models. The introduction at specific sites of amino acids which can be targets for specific chemical probes or cross-linking reagents, can allow the accessibility of those sites and their involvement in protein–protein interactions to be evaluated. The spatial constraints identified by this approach allow a structural map to be developed which more reliably reflects the authentic conformation of the molecule *in vivo.*

---

*This chapter is reproduced from *Alfred Benzon Symposium No. 39* (1995) with permission from Munksgaards Forlag, Copenhagen, Denmark.

A molecular model for the 16 kDa proteolipid

The diverse transport roles of the 16 kDa proteolipid. We have applied the approach based on molecular biology outlined above to the analysis of structure–function relationships in a 16 kDa integral membrane protein. This highly hydrophobic polypeptide is representative of a conserved class of proteins implicated in a variety of membrane transport processes (Finbow *et al.*, 1995). It is, for example, the membrane-bound component of the vacuolar H^+-translocating ATPase, in which it plays an essential role in H^+ conduction across the bilayer (Nelson, 1992). Related forms of the proteolipid may be involved in functions as diverse as regulation of neurotransmitter release (Birman *et al.*, 1990), in drug resistance (Shih *et al.*, 1990) and metal ion transport (Li *et al.*, 1994).

The 16 kDa proteolipid which provides our model experimental system is isolated from the lobster *Nephrops* in the form of paracrystalline arrays of protein particles in paired membrane structures. These appear reminiscent of gap junctions of higher animals (Finbow *et al.*, 1984). This, together with cell biological data, has been interpreted as suggesting another role for the proteolipid, namely in non-specific movement of solutes between the cytoplasmic phases of adjoining cells, and therefore, implicitly, in cell–cell communication.

The *Nephrops* hepatopancreas 16 kDa proteolipid provides a particularly good system for membrane protein structure–function studies; the primary structure of the protein is well-characterized and secondary structure predictions are supported by preliminary protein chemical, immunological and Fourier transform infrared (FTIR) spectroscopy studies (Holzenburg *et al.*, 1993). Engineering and subsequent study of the protein is relatively straightforward because it can be expressed at relatively high levels in a *Saccharomyces* system (Harrison *et al.*, 1994). Finally, the gap junction-like structures isolated directly from the lobster are amenable to electron microscopy, providing independent data concerning the macromolecular organization of the proteolipid protein.

Electron microscopy of the 16 kDa proteolipid. Electron microscopy images of the *Nephrops* gap junction-like complexes have been obtained with the help of Andreas Holzenburg at the University of Leeds (Holzenburg *et al.*, 1993). Enhancement by optical filtering and rotational averaging reveals each unit of the paracrystalline gap junctional array to be a regular hexamer of protein surrounding a stain-penetrable pore of diameter approximately 1.5 nm (*Figure 1*). From considerations of protein density, a mass for each unit of $\approx$100 kDa can be estimated, consistent with a composition of six 16 kDa proteolipids (Holzenburg *et al.*, 1993). The quaternary structure of the proteolipid is likely, therefore, to be a hexamer of individual polypeptides surrounding a staining central pore.

Primary structure, membrane topology and folding. The 16 kDa proteolipid is predicted to assemble in the lipid bilayer with four anti-parallel α-helices (FTIR

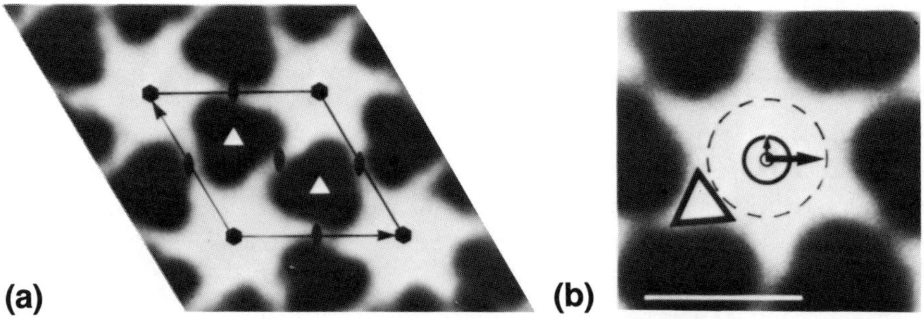

Figure 1. Electron microscopy of gap junction-like structures from *Nephrops*. (a) Reconstructed projection of the gap junctional unit cell after optical filtering, rotational averaging with centres of two-, three-, and six-fold symmetry indicated. Paracrystalline arrays were negatively stained, therefore protein appears white. (b) Reconstructed projection of a single hexameric unit, after optical filtering and six-fold averaging. The central pore indicated by the small arrow has a radius of 0.75 nm, whereas the radius of the surrounding ring measures 1.9 nm from the centre of the complex. The scale bar represents 5 nm. Reproduced from Holzenburg *et al.* (1993) by permission of the European Journal of Biochemistry.

suggests that the protein is >60% α-helix) connected by relatively short extramembranous loops (*Figure 2a*). In the case of the proteolipid which functions as a vacuolar H^+-ATPase subunit, these loops would be on the cytoplasmic surface of the membrane and would be expected to interact with the more water soluble subunits of the whole complex. Consistent with this role, they are the most highly conserved, in contrast to the N- and C-termini and 2/3 loop, all of which are on the other intraluminal face of the membrane.

The transmembrane helices of the proteolipid have been modelled as a tightly-packed classical four-helical bundle, and by mapping the hydrophobicity of each helical face, the relative orientation of individual helices within the hexameric proteolipid complex has been predicted (Finbow *et al.*, 1993) (*Figure 2b*) It is far from likely, however, that the relative orientation of the four helices in the bundle will be exactly the same as seen in the equivalent motifs of various water soluble proteins, but such a representation does allow useful residue dispositions to be portrayed. Helix 1 is devoid of any continuously hydrophobic faces, and is predicted to line the putative pore shown to be present by electron microscopy. Helix 3 has opposing hydrophobic and hydrophilic faces, whereas helices 2 and 4 show hydrophobicity broadly throughout. The hydrophilic face of helix 3 is predicted to face into the four-helical bundle. The opposing hydrophobic face of helix 3, which contains an interesting and pronounced glycine 'stripe', along with faces of helix 2 and helix 4, presents a continuously hydrophobic surface which is predicted to contact the lipid phase (*Figure 2b*).

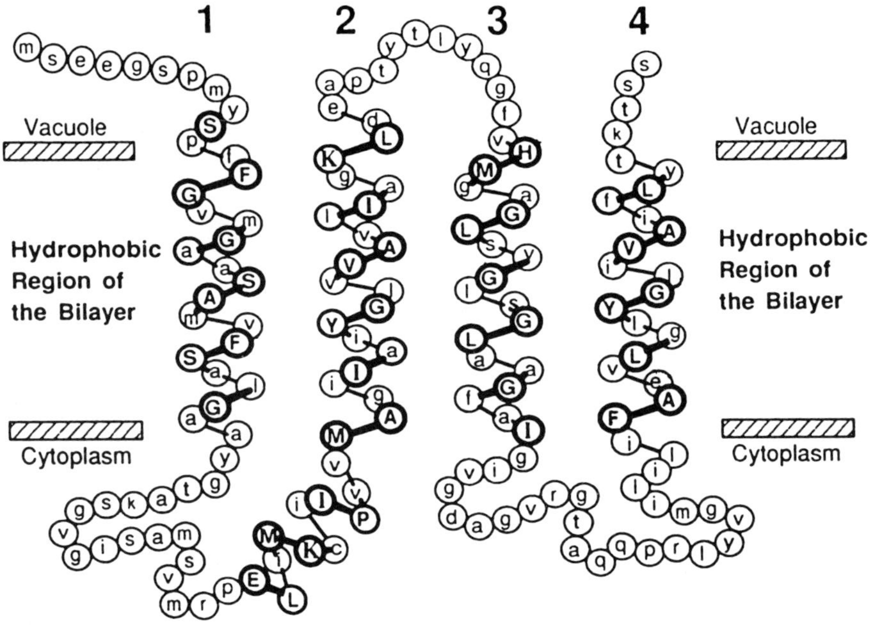
1
2
3
4
Vacuole
Vacuole
Hydrophobic Region of the Bilayer
Hydrophobic Region of the Bilayer
Cytoplasm
Cytoplasm

(a)

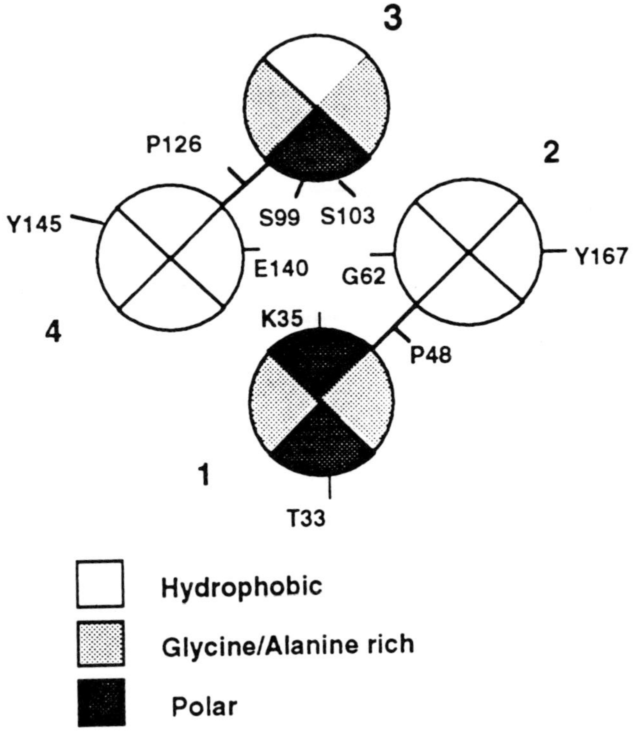
3
P126
2
Y145
S99
S103
E140
G62
Y167
K35
4
P48
1
T33
Hydrophobic
Glycine/Alanine rich
Polar

(b)

Helix 4 of the proteolipid contains a conserved glutamate residue. The presence of an acidic residue within a transmembrane helix is characteristic of virtually all proteolipids associated with F- or V-type ion-motive ATPases, and is crucial for ion transport function (Noumi *et al.*, 1991). The glutamate residue of the *Nephrops* proteolipid can be directed into the centre of four-helical bundle if it is assumed that such a charged residue could not, for purely thermodynamic reasons, point into the lipid phase. However, this residue is modified by the lipophilic reagent dicyclohexylcarbodiimide (DCCD) in a reaction which forms a stable product only in the absence of water, suggesting that it is accessible from a hydrophobic environment such as the lipid phase (Noumi *et al.*, 1991). There are also no strong thermodynamic reasons for discounting the possibilities of a *protonated* glutamate being present at the lipid–protein interface. The position of this key residue with respect to the four-helical bundle is discussed further below.

Computer-graphic modelling of the 16 kDa proteolipid. To build a model of the *Nephrops* proteolipid, the aligned sequences of related 16 kDa proteolipids were initially analysed by eight different secondary structure prediction, hydrophobicity profile or hydrophobicity moment algorithms (Finbow *et al.*, 1992).

To generate the three-dimensional model, (*Figure 3*, see page 112), techniques similar to those used in (Hecht *et al.*, 1990) were used. The four-helical segments which were determined by structural prediction/hydrophobicity profile were assembled such that polar or charged residue accessibility was minimized, and hydrophobic surfaces optimized (see *Figure 2b*). A 9–10 Å spacing and tilts of +10° to +15° were imposed, as commonly found in natural four-helical bundles. Bulky hydrophobic side chains were positioned so that they packed against one another on the surfaces of neighbouring helices. The conformations of the extramembranous loops were derived from a combination of homology searches in the Brookhaven structural databases and molecular mechanics optimization.

Application of six-fold symmetry to a single four-helical bundle generates a model which is very similar in both dimensions and conformation to the electron microscopy images of the *Nephrops* gap junction-like units (*Figure 1*).

Figure 2. Transmembrane topology, helix packing in the *Nephrops* 16 kDa proteolipid (a) Predicted disposition of the 16 kDa proteolipid within the vacuolar membrane. The polypeptide is predicted to fold with four transmembrane α-helices. Note that the helical segments are frequently bounded by charged or polar residues. With the exception of the critical glutamate residue in helix 4 (Glu-140), charged residues are restricted to extramembranous regions. (b) Proposed packing of transmembrane helices. Only helix 1 is entirely devoid of completely hydrophobic faces, and is proposed to have no contact with the lipid phase. Helix 3 has a single face containing polar residues, proposed to be directed into the centre of a four-helical bundle. Helices 2 and 4 are positioned to minimize exposure of the few polar residues. The outer face of helix 3 together with those of helices 2 and 4 provide a continuously hydrophobic face in contact with the lipid phase.

2. Testing the molecular model of the 16 kDa proteolipid

Expression of the Nephrops proteolipid in S. cerevisiae

Reasonably high level expression of the Nephrops proteolipid is a prerequisite for the protein chemistry and activity measurements which will provide a more detailed picture of the structure–function relationships of the protein. For the purposes of expression, we have successfully exploited a yeast strain in which the VMA3 gene, encoding the endogenous proteolipid V-ATPase subunit, has been disrupted (Nelson and Nelson, 1990). Because of the highly conserved nature of the 16 kDa proteolipids, transformation of this *vma*3 yeast strain with a cDNA encoding the *Nephrops* protein results in functional substitution of the yeast protein. The *Nephrops* 16 kDa proteolipid is incorporated into a hybrid V-ATPase, containing yeast water-soluble subunits, giving complete restoration of V-ATPase function and complementation of the *vma*3 mutation (Harrison *et al.*, 1994) A conditional lethal phenotype is diagnostic for mutations in genes for V-ATPase subunits, characterized by sensitivity to growth at pH >7 and to high Ca^{2+} concentrations (Nelson and Nelson 1990). Growth at mildly acid pH 5–6 is relatively unaffected. The recovery of V-ATPase function in the vma3 yeast strain by expression of the *Nephrops* proteolipid is manifested as an abolition of the conditional lethal phenotype and restoration of growth at pH 7.5. This system, therefore, provides not only high levels of expression, with the expressed protein targeted to a membrane system suitable for protein chemical studies, but also a convenient means of assessing functional effects of site-direct mutations.

Cysteine-replacement mutagenesis

The reactive and characteristic nature of the thiol group means that cysteine side-chains are particularly good targets for protein chemical studies aimed at mapping the relative positions of individual residues within a protein. A number of strategies which exploit this reactivity have successfully been applied to structure–function studies of a variety of membrane proteins. These strategies generally involve the introduction by site-directed mutagenesis of cysteine residues at specific sites (Akabas *et al.*, 1992; Altenbach *et al.*, 1990; Milligan *et al.*, 1988; Pakula and Simon, 1992). These substituted cysteines can subsequently be probed with specific reagents of varying size or chemical character, allowing discrimination between residues which are accessible from the aqueous phase, or which are adjacent to the lipid phase.

We have developed this approach to provide a strategy for testing the published molecular model for the *Nephrops* 16 kDa proteolipid in experiments designed to test specific predictions concerning the position and orientation of segments of the proteolipid (Jones *et al.*, 1995).

Probing cysteine residues with dye-substituted maleimides

The *Nephrops* 16 kDa proteolipid has a single cysteine residue (Cys-54) see *Figure 2a*) which we have mutated to serine to provide a fully active cysteine-free form of proteolipid. This in turn has served as a template for the individual mutation of eight consecutive residues of helix 1 to cysteines (from Phe-24 toTyr-31; see *Figure 2a*), providing a series of mutants mapping out two complete turns of an α-helix. Additional cysteine replacements were made in helix 3 (Ser-103, Leu-105), near the exposed N-terminus (Ser-6) and in the H1–H2 extramembranous loop (Ser-44).

Each mutant form of the proteolipid has been expressed in the *S. cerevisiae vma*3 strain, and the vacuolar membrane V-ATPase function characterized in each case. Cysteine replacement of residues Phe-24, Ser-25 and Tyr-31 resulted in V-ATPase dysfunction, characterized by slow growth at pH 7.5 (relative to the non-mutated strain) and high K_m for ATP hydrolysis. Replacement of Gly-28 resulted in no growth at pH 7.5 and a very high K_m for the V-ATPase. Other mutations had no effect either on growth of the yeast strain or on enzymological properties of the respective hybrid V-ATPases *in vitro* (*Table 1*).

In order to probe accessibility of cysteine residues to the aqueous phase we have devised a procedure for screening for modification by the water-soluble reagent fluorescein-5-maleimide (F-M). Vacuoles isolated from each yeast strain were incubated with the reagent and the vacuolar membrane proteins subjected to sodium dodecyl sulphate–polyacrylamide gel electrophoresis (SDS–PAGE) and immunoblotting (*Figure 4*). Modification by F-M results in a shift in mass of the proteolipid on SDS–PAGE, a property which we have used to test each cysteine replacement mutant for reactivity to the maleimide reagent. Replacement of Ala-26, Ala-29, Ala-30, Tyr-31, Ser-44 and Ser-6 all gave cysteine residues which could be modified by F-M, indicating their accessibility from the aqueous phase.

Similarly, accessibility from the lipid phase was tested with the hydrophobic reagent benzophenone-4-maleimide (*Figure 4*). The only residue so far tested showing reactivity with this probe was a cysteine replacing Leu-105 of helix 3 (see *Figure 2a*), suggesting proximity of this helix to the lipid bilayer (*Figure 2b*). None of the residues in helix 1 could be modified by this hydrophobic reagent.

Functional and dye-maleimide sensitivity data are summarized in *Table 1* and *Figure 5*.

However, a better appreciation of the data is gained when the distribution of functionally sensitive and fluorescein maleimide-accessible residues of helix 1 are mapped on a helical wheel projection (*Figure 6*)

From *Figure 6* it is clear that residues whose mutation to cysteine produces dysfunction and those which are not sensitive are clustered onto opposite faces of

Table 1. Summary of data for cysteine-replacement mutants of the *Nephrops* proteolipid

Strain	Growth pH 7.5[a]	K_m	F-5-M labelling[b]	B-4-M labelling[b]	Disulphide X-Ligand[c]
Nephrops 16K	+	0.4	±	–	–
C54S	+	0.5	–	–	–
Helix 1:					
F24C	±	1.2	–	–	–
S25C	±	0.9	–	–	–
A26C	+	0.5	+	–	+
L27C	+	0.5	–	–	–
G28C	–	3.0	–	–	–
A29C	+	0.7	+	–	–
A30C	+	0.4	+	–	–
Y31C	±	1.6	+	–	–
Extramembranous:					
S6C	+	0.4	+	–	+
S44C	+	0.5	+	–	+
Helix 3:					
G101C	–	0.3	–	–	–
S103C	+	0.8		–	–
L105C	+	0.4	–	+	–
Double mutants:					
S25C/L27C	+	n.d.	n.d.	n.d.	+
A26C/A29C	–	n.d.	n.d.	n.d.	–
L27C/A29C	+	n.d.	n.d.	n.d.	+

[a]Growth at pH 7.5 is expressed relative to the rate of growth of the *S. cerevisiae* strain expressing the non-mutated form of *Nephrops* proteolipid. Values of K_m for ATP hydrolysis by hybrid V-ATPases *in vitro* are expressed as mM ATP.
[b]Accessibility to dye-maleimides was determined by mass shift electrophoresis (see legend to *Figure 5).*
[c]Disulphide cross-linking was monitored as formation of covalently-linked dimers on non-reducing SDS–PAGE.

helix 1. Residues insensitive to mutation are also largely accessible to the water-soluble reagent F-M (indicated by asterisks), implying that this face of the helix lines the central pore which is evident in electron micrographs. The mutation-sensitive face of helix 1 is likely therefore to be oriented into the centre of the four-helical bundle. Analysis of aligned sequences of related proteolipids by the programme PERSCAN V7.0 (Donnelly *et al.*, 1993) supports this assumption: the sensitive face of helix 1 is also the most conserved, in accord with general rules which suggest that conserved faces tend to be internal within folded proteins (Donnelly *et al.*, 1993; Komiya *et al.*, 1988). The observation that mutagenesis of the pore-lining face of helix 1 exerts no functional effects has implications for the H^+-translocating function of the proteolipid, and this is discussed below.

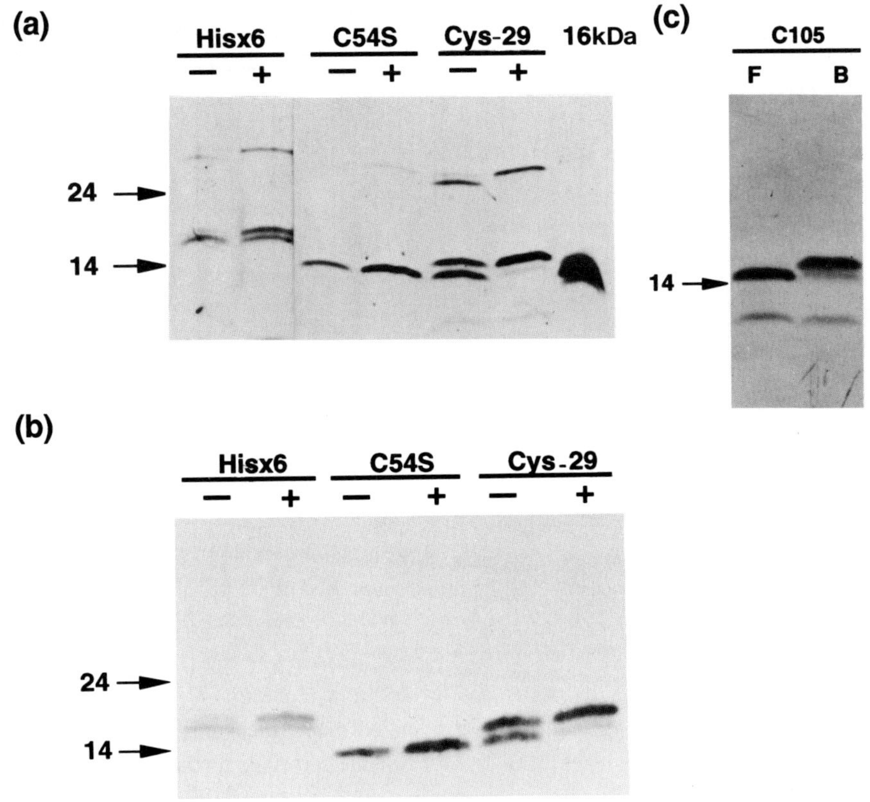

Figure 4. Estimation of dye-maleimide modification by mass-shift electrophoresis. Vacuolar membranes isolated from *S. cerevisiae* strains expressing different forms of *Nephrops* proteolipid were incubated in the presence of 100 μM dithiothreitol, then probed with 5 mM fluorescein-5-maleimide (F-M) for 60 minutes at 30°C, in the presence (+) or absence (–) of SDS. Hisx6: non-mutated *Nephrops* proteolipid with C-terminal polyhistidine affinity tag (Harrison *et al.*, 1994). C54S: cysteine-free form of *Nephrops* proteolipid. Cys-29; cysteine replacement mutant at residue Ala-29. 16 kDa indicates gap junctional protein isolated from the lobster. The reaction is terminated by addition of excess 2-mercaptoethanol, the proteolipid extracted with chloroform:methanol (2:1) at 4°C (Arai *et al.*, 1987). Extracted proteolipid was analysed by SDS–PAGE followed by (a) silver staining or (b) immunoblotting. Cysteine residues introduced by replacement mutagenesis (Cys-29) or present in the normal proteolipid (Cys-54; see *Figure 2(a)*) modified by F-M gave rise to a mass increase on SDS–PAGE. The proportion of proteolipid modified was enhanced by denaturation of the protein by SDS. Note also the mass increase caused by the addition of the C-terminal polyhistidine tag. In the silver-stained gel (a), an additional higher mass polypeptide which is chloroform:methanol soluble but which is unrelated to the proteolipid also shows F-M modification. This procedure was used to screen the cysteine replacement as mutants detailed in the text. (c) A similar, although smaller, mass shift was also observed after modification of the L105C mutant proteolipid by benzophenone-4-maleimide.

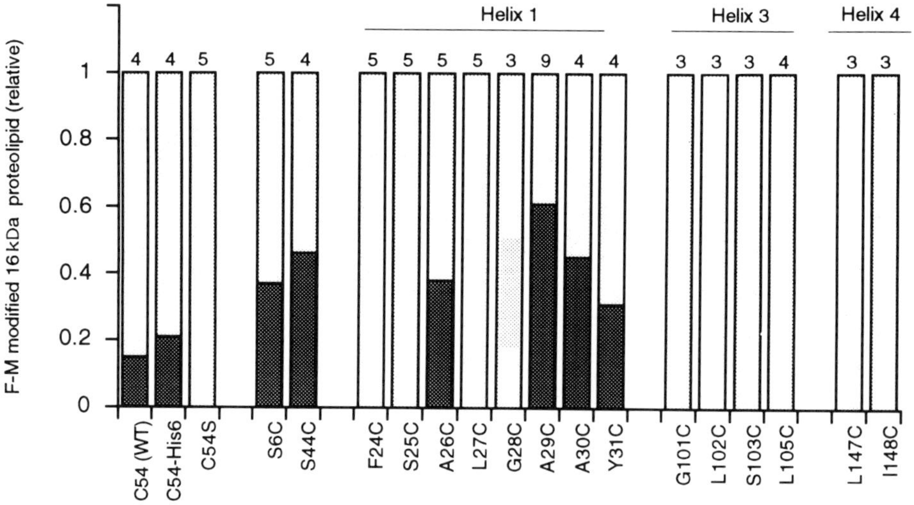

Figure 5. Summary of residue modification by F-M. Using the conditions described in *Figure 4*, the relative proportions of F-M modified (dark shading), unmodified (light shading) protein were determined for each cysteine mutant by laser densitometry of the appropriate silver stained polyacrylamide gels.

The third transmembrane helix of the proteolipid contains a very distinctive structural feature, an extended series of glycine residues (clearly seen in *Figure 2*). These glycine residues occur at three or four residue intervals throughout the helix. The precise function of this feature remains unclear, but it is certainly highly conserved and particularly sensitive to mutagenic substitutions (Kim *et al.*, unpublished; Noumi *et al.*, 1991). In structural terms, a series of residues lacking side chains would be expected to produce a concave feature on one helical face. This 'smooth' strip of residues might be essential for the tight packing of the four-helical bundle, and would (according to the structural model) be in contact with helix 4 and perhaps partially accessible to the lipid phase. This is consistent with assigned positions of helix 3 residues neighbouring the glycine strip. A cysteine replacing Leu-105 is reactive towards benzophenone maleimide and likely, therefore, to be lipid-exposed, whereas a cysteine replacement of Ser-103 provides a thiol inaccessible to both thiol F-M and B-M and produces a V-ATPase with functional properties characteristic of mutations within the core of the four helical bundle (high K_m, slow growth at pH 7.5).

Probing helix–helix interactions by disulphide cross-linking

A second strategy, which is proving even more useful in the study of transmembrane helix orientation and packing, relies on the capacity of pairs of cysteine residues to form disulphide bridges in the native active protein. By analysing the propensity of mutagenically introduced cysteine residue to form

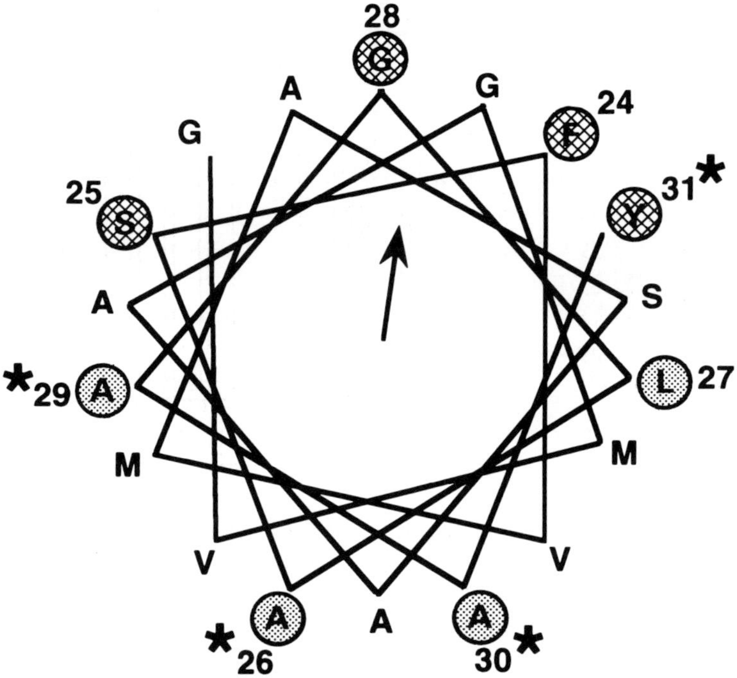

Figure 6. Helical wheel projection of transmembrane helix 1. Helix 1 of the *Nephrops* 16 kDa proteolipid, plotted in a clockwise direction viewed from the C-terminal (cytoplasmic) end. Residues changed by cysteine replacement mutagenesis are ringed . (▩) Residues functionally sensitive to mutation, giving rise to high K_m for ATP for the hybrid vacuolar H^+ATPase and reduced growth at pH 7.5. (░) Residues not affecting V-ATPase function. Residues accessible to fluorescein maleimide are indicated by asterisks. The arrow indicates the most conserved face of the helix, determined by the programme PERSCAN (D. Donnelly, this laboratory) (Donnelly *et al.*, 1993).

cross-links with cysteine residues in neighbouring helices, data can be gathered concerning the identities of interacting residues and the orientation of helices with respect to one other.

Cross-linking between cysteine residues may be either inter- or intra-molecular, with covalent associations being detected by mass shift on SDS–PAGE run under non-reducing conditions. Intermolecular associations give rise to dimeric species, manifested as an increase in molecular mass, whereas intra-molecular cross-links generally give rise to an apparent decrease in mass.

It is reasonable to assume that the apparent six-fold symmetry incorporated into the molecular model for the proteolipid genuinely reflects the native structure of the protein *in vivo* (Holzenburg *et al.*, 1993). Assuming also that general rules for helical packing are observed, it follows that only the transmembrane helix which is proximal to the central pore can interact directly with the same helix of neighbouring polypeptides. The F-M labelling and mutagenesis data strongly indicate that helix 1 holds this location. We have monitored cross-linking between cysteine residues in our replacement mutants and shown that covalently-linked dimers of proteolipid are generated in the case of both S6C and S44C replacement mutants, positioned at the N- and C-terminal ends of helix 1 respectively (*Figures 2a* and *7*) (Jones *et al.*, 1995). The only cysteine replacement within helix 1 giving rise to cross-linking and dimerization was at position 26 (A26C) (Jones *et al.*, 1995). Because of considerations of symmetry and spacing, cross-linking between residues at positions 6, 26 or 44 must indicate that helices 1 of each proteolipid are adjacent to each other within the hexameric complex (*Figure 7*).

In the case of the cysteine at position 26, the regular six-fold symmetry of the hexameric proteolipid complex means that the respective Cys-26 residues of neighbouring polypeptides would be closest to each other if they pointed directly into the central pore of the complex (*Figure 7*). This inference is in turn supported by the F-M labelling data which showed accessibility of this residue but not of residues 25 or 27.

The cross-linking data are, therefore, not only consistent with a model in which helix 1 lines the central pore, but also provide independent support for the proposed orientation of helix 1 within the four-helical bundle which we propose to be the basic folded unit of the proteolipid complex

If the structural model is correct with respect to helix 1, there are further structural implications which can be tested. Residues at positions 25 and 29 of one polypeptide would be close to residue 27 of a neighbouring polypeptide. Neither of these residues, however, would have proximity to a cysteine at position 26 in a neighbouring helix (*Figure 7*).

Examination of helical models suggest that cysteine side chains at positions 25 and 29 could be ~3 Å (from the side chain of a cysteine at position 27 in an adjacent helix, sufficiently close to allow formation of disulphide cross-links (Pakula and Simon, 1992). We have tested this by constructing proteolipids containing pairs of cysteine residues separated by ~200° (S25C/.L27C, L27C/A29C) and ~60° (A26C/A29C) (see *Figure 6*). Current data do indeed suggest that cross-linking can occur between a Cys-27 residues and either Cys-25 or Cys-29 of a neighbouring helix (Murray *et al.*, unpublished data).

3. Functional implications of proteolipid structure

Pathways of ion movement

The unambiguous positioning of helix 1 at the centre of the proteolipid hexamer, and the orientation of mutationally-insensitive helix 1 residues into the central pore

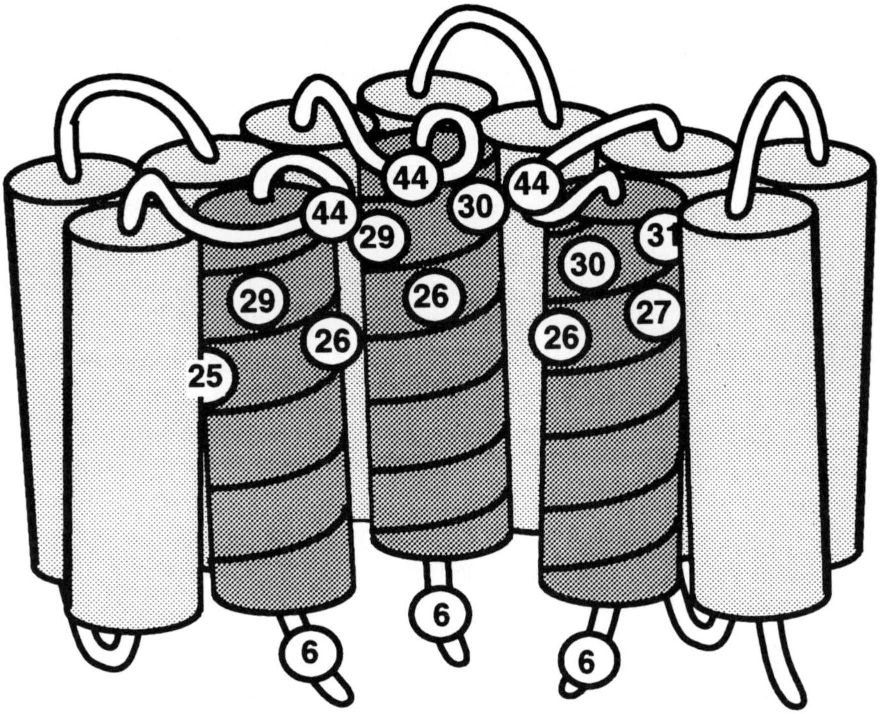

Figure 7. Proposed orientation of helix 1 within the hexameric proteolipid complex. Spatial constraints dictated by disulphide cross-linking between cysteines introduced at residues 6, 26 or 44 indicate proximity of helix 1 (dark shading) to the same helix of neighbouring polypeptides. Folding of the loop between helix 1 and 2 is drawn to imply proximity between Cys-44 residues. Cysteines at residue positions 26, 29, 30, 31, 6, 44 are all accessible to F-M. Accessibility of Cys-31 is likely to be due to its proximity to the cytoplasmic phase.

indicates that this pore is unlikely to be a specificity-determining route for proton movement. Indeed, in the isolated V_0 complex such a central pore is likely to remain closed (Harrison *et al.*, 1994; Zhang *et al.*, 1992), perhaps as a consequence of the presence of other subunits (Bauerle *et al.*, 1993). The permeability which is observed when the 16 kDa proteolipid is reconstituted on its own (Sun *et al.*, 1987; Zhang *et al.*, 1994) may reflect the existence of an open central pore and a complex functionally analogous to the gap junction-like structures found in arthropods. The non-specific measurement of relatively large molecules which occurs during gap junctional communication would require a large channel, whereas movement of a proton could occur by exerting along a specific pathway.

Certainly, in the case of the V-ATPase is seems that the pathway of proton movement is contained within some other structural feature. The pronounced functional effects of mutating internal residues could imply that the centre of the four helical bundle may contain this pathway, particularly if the structural model is correct with respect to the internal orientation of the critical Glu-140 residue. The presence of six separate proton translocating pathways is consistent with ATP/H^+ stoichiometries. However, there must be a high degree of co-operativity between these pathways, since elimination of only one of six potential pathways by reaction with DCCD eliminates proton translocation completely (Arai *et al.*, 1987).

The face of helix 4 which contains Glu-140 is the most conserved in this helix, consistent with an internal orientation. A caveat must be attached to this observation, however. Because of its inherent assumptions, computer-assisted alignment and conservation analysis may not readily accommodate the possibility that a lipid-exposed face may have a high degree of functionality and would, therefore, be highly conserved. In fact helix 4 is highly conserved throughout, and it is equally possible that the conserved face with Glu-140 performs a specialized function oriented more towards the bilayer, consistent with its accessibility to DCCD. Indeed, preliminary data on the ability of Ser-103-Cys to suppress a dysfunctional Ile-148-Cys mutation would tend to indicate a bilayer orientation for Glu-140 (Y.-I. Kim, unpublished data).

The structural observations, coupled to existing knowledge concerning inhibitor binding, ion conductance properties and substrate stoichiometries can support a hypothesis that the mechanism of V-ATPase action involves the functional interaction of lipid-exposed regions of the proteolipid with other V-ATPase subunits. Proton translocation might occur in a cyclical mechanism involving each individual Glu-140 in turn. It is easy to see how such a cycle would be blocked by a single DCCD modification.

Further refinements to the structural model

Although structural information is beginning to emerge, much more data are required if a complete picture of the proteolipid structure is to be obtained. Suppressor mutations in the yeast proteolipid may provide clues concerning interfacing helices (Supek, 1994) which can be used to guide cysteine replacements. Protein engineering of double cysteine mutant forms of proteolipid or *in vitro* expression of single cysteine mutants, can produce the cross-linking data with which we can assign specific intra- and inter-molecular associations.

The use of cysteine-specific probes provides information concerning accessibility within the protein complex and defines the residues at the membrane boundary. The development of new probes may enhance the definition which can be obtained. This approach may also provide a degree of functional information. Labelling with cysteine probes, for example, may be conformationally sensitive and allow analysis of changes which occur during assembly/disassembly of the V-ATPase, or even more excitingly, during cycles of catalysis.

Acknowledgements

This work was funded by grants from the BBSRC to JBCF and MEF, by the Wellcome Trust to PCJ and The British Council to Y-IK, to whom we extend our grateful thanks.

References

Akabas, MH, Stauffer DA, Xu M, Karlin A. (1992) Acetylcholine receptor channel structure probed in cysteine-substitution mutants. *Science*, **258**: 307–310.

Altenbach C, Marti T, Khorana HG, Hubbel WL. (1990) Transmembrane protein structure: spin labeling of bacteriorhodopsin mutants. *Science*, **248**: 1088–1094.

Arai H, Berne M, Forgac M. (1987) Inhibition of coated vesicle proton pump, labeling of a 17,000 dalton polypeptide by *N,N′*-dicyclohexyl carbodiimide. *J. Biol. Chem.* **262**: 11006–11011.

Bauerle, C, Ho, MN, Lindorfer, MA, Stevens, TH. (1993) *The Saccharomyces cerevisiae* VMA6 gene encodes the 36 kDa submit of the vacuolar H^+-ATPase membrane sector. *J. Biol. Chem.* **268**: 12749–12757.

Birman S, Meunier F-M, Lesbats B, le Caer J-P, Rossier J, Isräel MA. (1990) 5 kDa proteolipid found in mediatophore preparations from *Torpedo* electric organ presents high sequence homology with the chromaffin granule protonophore. *FEBS Lett.* **261**: 303–306.

Donnelly, D, Overington, JP, Ruffle, SV, Nugent, JHA, Blundell, TL. (1993) Modelling α-helical transmembrane domains: The calculation, use of substitution tables for lipid facing residues. *Protein Sci.* **2**: 55–70.

Finbow ME, Buultjens TEJ, Lane NJ, Shuttleworth J, Pitts JD. (1984) Isolation, characterisation of arthropod gap junctions. *EMBO J.* **3**: 2271–2278.

Finbow ME, Eliopolous EE, Jackson PJ, Keen JN, Meagher L, Thompson P, Jones PC, Findlay JBC. (1992) Structure of the 16 kDa integral membrane protein that has identity to the putative proton channel of the vacuolar H^+-ATPase. *Protein Eng.* **5**: 7–15.

Finbow ME, Jones PC, Harrison MA. (1995) Bioessays, Ductin – a proton pump component, a gap junction channel, a neurotransmitter release channel. *Bioessays*, **17**: 244–247.

Harrison, MA, Jones, PC, Kim, Y-I, Finbow, ME, Findlay, JBC. (1994) Functional properties of a hybrid vacuolar H^+-ATPase. *Eur. J. Biochem.* **221**: 111–120.

Hecht MH, Richardson JS, Richardson DD, Ogden RC. (1990) *De novo* design, expression, characterisation of Felix: a four-helix bundle protein of native-like sequence. *Science*, **249**: 884–891.

Holzenburg A, Jones PC, Franklin T, Pali T, Heimburg T, Marsh D, Findlay JBC, Finbow ME. (1993) Evidence for a common structure for a class of membrane channels. *Eur. J. Biochem.* **213**: 21–30.

Jones PC, Harrison MA, Kim Y-I, Finbow ME, Findlay JBC. (1995) The first putative transmembrane helix of the 16 kDa proteolipid lines a pore I the V_0 sector of the vacuolar H^+ATPase. *Biochem. J.* **312**: 739–749.

Komiya H, Yeates TO, Rees DC, Allen SP, Feyer, G. (1988) Structure of the reaction center from Rhodobacter sphaeroides R-26: symmetry relations, sequence comparisons between different species. *Proc. Natl Acad. Sci. USA*, **85**: 9012–9016.

Li, C-Y, Watkins, JA, Glass, J. (1994) The H^+-ATPase from reticulocyte endosomes reconstituted into liposomes acts as an iron transporter. *J. Biol. Chem.* **269**: 10242–10246.

Milligan, DL, Koshl and, DE Jr. (1988) Site-directed cross-linking. Establishing the dimeric structure of the aspartate receptor of bacterial chemotaxis. *J. Biol. Chem.* **263**: 6268–6275.

Nelson H, Nelson N. (1990) Disruption of genes encoding subunits of the yeast vacuolar H^+ ATPase causes conditional lethality. *Proc. Natl Acad. Sci. USA*, **87**: 3503–3507.

Nelson N. (1992) Organellar proton ATPases. *Curr Opin Cell Biol.* **4**: 654–660.

Noumi, T, Beltrán, C, Nelson, H, Nelson, N. (1991) Mutational analysis of yeast vacuolar H^+-ATPase. *Proc. Natl Acad. Sci. USA,* **88**: 1938–1942.

Pakula AA, Simon MI. (1992) Determination of transmembrane protein structure by disulfide crosslinking: the *E. coli Tar* receptor. *Proc. Natl Acad. Sci. USA,* **89**: 4144–4149.

Shih C-K, Kwong J, Montalvo E, Neff N. (1990) Expression of a proteolipid gene from a high copy number plasmid confers trifluoperazine resistance in *Saccharomyces cerevisiae. Mol. Cell Biol.* **103**: 3397–3404.

Sun S-Z, Xie X-S, Stone DK. (1990) Isolation, reconstitution of the dicyclohexylcarbodiimide-sensitive proton pore of clathrin-coated vesicle proton translocating complex. *J. Biol. Chem.* **262**: 14790–14794.

Supek F, Supekova L, Nelson N. (1994) Features of vacuolar H^+-ATPase revealed by yeast suppressor mutants. *J. Biol. Chem.* **269**: 26479–26485.

Zhang J, Myers M, Forgac M. (1992) Characterization of the V_0 domain of the coated vesicle H^+-ATPase. *J. Biol. Chem.* **267**: 9773–9778.

Zhang J, Feng Y, Forgac M. (1994) Proton conduction, bafilomycin binding by the V_0 domain of the coated vesicle V-ATPase. *J. Biol. Chem.* **269**: 23518–23523.

Index

ORDERING DETAILS

Main address for orders

BIOS Scientific Publishers Ltd
9 Newtec Place, Magdalen Road,
Oxford OX4 1RE, UK
Tel: +44 1865 726286
Fax: +44 1865 246823

Australia and New Zealand
DA Information Services
648 Whitehorse Road, Mitcham, Victoria 3132, Australia
Tel: (03) 9210 7777
Fax: (03) 9210 7788

India
Viva Books Private Ltd
4325/3 Ansari Road, Daryaganj, New Delhi 110 002, India
Tel: 11 3283121
Fax: 11 3267224

Singapore and South East Asia
(Brunei, Hong Kong, Indonesia, Korea, Malaysia, the Philippines, Singapore, Taiwan, and Thailand)
Toppan Company (S) PTE Ltd
38 Liu Fang Road, Jurong, Singapore 2262
Tel: (265) 6666
Fax: (261) 7875

USA and Canada
Books International Inc
PO Box 605, Herndon, VA 22070, USA
Tel: (703) 435 7064
Fax: (703) 689 0660

Payment can be made by cheque or credit card (Visa/Mastercard, quoting number and expiry date). Alternatively, a *pro forma* invoice can be sent.

Prepaid orders must include £2.50/US$5.00 to cover postage and packing (two or more books sent post free)